URBAN-RURAL LINKS

Issues in Geography

HEINEMANN
EDUCATIONAL

JOHN WESTAWAY   PAT COLLARBONE
JANE CONNOLLY   CHRIS SCRIVENER

30424

Heinemann Educational,
a division of Heinemann Educational Books Ltd,
Halley Court, Jordan Hill, Oxford OX2 8EJ

OXFORD   LONDON   EDINBURGH
MELBOURNE   SYDNEY   AUCKLAND
IBADAN   NAIROBI   GABORONE
HARARE   KINGSTON   PORTSMOUTH (NH)
SINGAPORE   MADRID
ATHENS   BOLOGNA

© John Westaway, Pat Collarbone,
Jane Connolly and Chris Scrivener 1990

First published 1990

British Library Cataloguing in Publication Data

Urban – rural links. – (Issues in geography).
  1. Urbanisation. Secondary school texts
  I. Westaway, John, *1949* – II. Series
  307.7'6

ISBN 0 435 349228

*Designed and produced by
Gecko Limited, Bicester, Oxon*

Illustrated by Gecko Limited and Sally Launder

*Printed and bound in
Spain by Mateu Cromo*

Contents

Rural and urban changes

Rural = 'to do with the countryside' (Walker, A, *Basic Dictionary of Geography*, Bell and Hyman, 1981) 'characteristic of the country or country life' (*Longman Dictionary of Geography*, 1985)

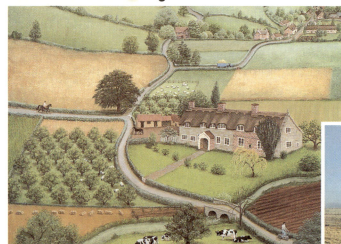

A 'Chocolate box' countryside

C Paddy field in Bali

B Wilderness

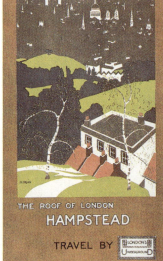

E London Underground advert

D Tropical rainforest in Ecuador

F Rural decay

4

G The Brixton riots, 1985

Urban = 'to do with a town or city' (Walker, A, *Basic Dictionary of Geography*, Bell and Hyman, 1981) 'characteristic of a town or city' (*Longman Dictionary of Geography*, 1985)

H Groveways, Brixton

I A shopping centre in Hong Kong

J Traffic congestion, Bangkok

K Central Park, New York, USA

L City night-life, Las Vegas

Image and reality (2)

What are your images?

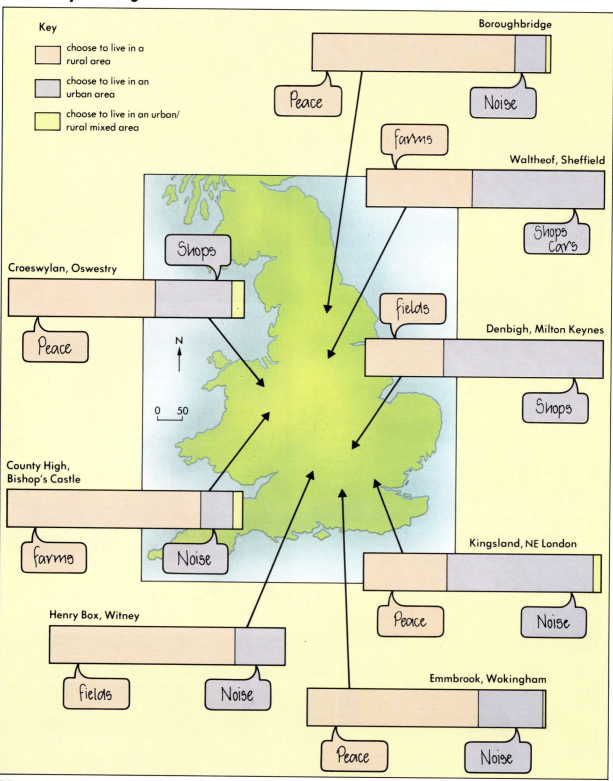

M Questionnaire results

This page shows the results of a questionnaire completed by pupils in eight schools in England and Wales. They were asked to write the words they associated with 'rural' and 'urban'. They were also asked to choose which place they would like to live in.

Where I'd like to live

These reasons for residential preference (liking one place better than another) were answers pupils gave to the questionnaire.

'I'd prefer to live in the countryside because . . .'

N Reasons for preferring to live in the countryside

'I'd prefer to live in the city because . . .'

O Reasons for preferring to live in the city

What is rural?

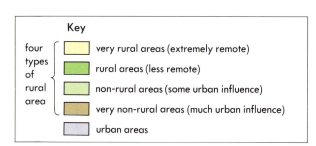

Key

four types of rural area
- very rural areas (extremely remote)
- rural areas (less remote)
- non-rural areas (some urban influence)
- very non-rural areas (much urban influence)
- urban areas

10 areas which are very non-rural	10 areas which are very rural
1 Whiston	i Newcastle Emlyn
2 Meriden	ii New Radnor
3 Warrington	iii Painscastle
4 Wokingham	iv Reeth
5 Highworth	v Machynlleth
6 Hartley Wintney	vi Knighton
7 Easthampstead	vii Lleyn
8 Grimsby	viii Llanfyllin
9 Wigan	ix Penllyn
10 Blaby	x Kington

P Very non-rural and rural districts

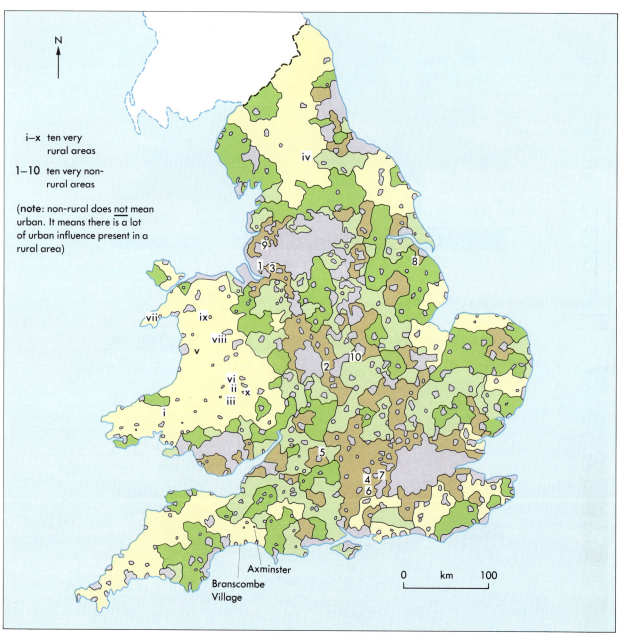

N

i–x ten very rural areas

1–10 ten very non-rural areas

(**note:** non-rural does <u>not</u> mean urban. It means there is a lot of urban influence present in a rural area)

Axminster

Branscombe Village

0   km   100

Q The four types of rural area and urban areas in England and Wales (*Cloke, P J, 'An Index of Rurality for England and Wales', Regional Studies, 11, 1977*)

8

County	Percentage	County	Percentage
Cornwall	26.5	Somerset	16.0
Borders	23.6	Avon	15.2
Dumfries and Galloway	22.4	Lincolnshire	14.9
Devon	21.4	Suffolk	14.8
Dyfed (excluding Llanelli)	20.9	Dorset	14.6
Gwynedd	19.5	Cambridgeshire	14.1
Tayside	18.3	West Sussex	13.7
East Sussex	18.1	Lancashire	13.7
Highland	17.7		
Isle of Wight	17.7	(Average for Great Britain = 10%)	
Hereford and Worcester	17.2		
North Yorkshire	17.1		

R Counties with a high percentage of low-paid male workers (*Thomas and Winyard, 1979*)

> I live in Greater London. I work in a bank's Head office. I earn £170·00 a week.

> I live in the South West. I am an Agricultural worker. I earn £129·00 a week

S Female workers (*Regional Trends, 23, 1988*)

T Gross weekly earnings (full-time) (*Social Trends, 18, 1988*)

U Rural transport

> I'm so worried, my husband needs regular hospital treatment and I don't know how we're going to get there.

V A rural transport problem

This woman from Branscombe Village, South Devon does not own a car. The hospital is in Axminster. (See Map Q page 8 for the location of these places.)

Conflict in the countryside

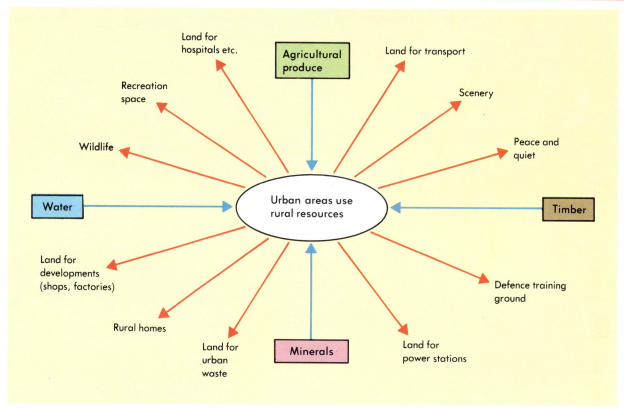

A The town uses the countryside as a resource

When was last trip made?	% of population
In last 7 days	24 frequent
Over 1 week, up to 4 weeks	29 occasional
Over 4 weeks, up to 3 months	11 occasional
Over 3 months, up to 6 months	10 occasional
Over 6 months, up to 12 months	10 rarely
Longer than a year	11 rarely
Never	5 rarely

B How frequently people visit the countryside (*Household Surveys for the Countryside Commission, Summer 1986*)

- On a summer Sunday, up to 18 million trips are made into the countryside, involving two fifths of the entire population of England and Wales.

- In the course of a year, 85 per cent of us visit the countryside at some stage.

- Two other important results from the survey are that nearly half of all countryside outings are within 10 miles of home, and three-quarters are made by car.

D Managed countryside = country parks, stately homes, etc (*Household Surveys for the Countryside Commission, Summer 1986*)

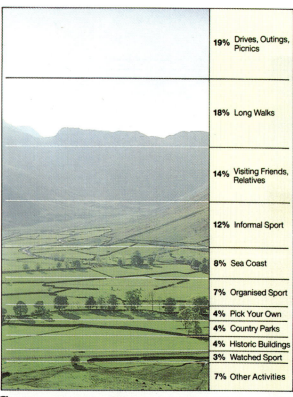

C Countryside recreation activities, 1984 survey (*Countryside Commission News, 18, December 1985*)

Rural resources: who gains, who loses?

E An artist's impression of the exploitation of rural resources (*Worldwide Fund for Nature*)

Minerals and the countryside

F Opencast mining is on the increase in many counties in England and Wales

CPRE IN ACTION

The Council for the Protection of Rural England (CPRE) presented a strong case to the House of Commons Energy Select Committee, which concluded that opencast coal output should be reduced.

At public inquiries in the North East, CPRE and local authorities have repeatedly convinced Inspectors that proposals for opencasting on greenfield sites should be refused because of the environmental damage they would cause.

G Extract from the *CPRE Annual report*, 1987

Urbanization (1)

Settlement: size, function and change

A   Road map of northern France, scale 1:200 000 (Reproduced with the permission of *Michelin, from Map no. 55*)

C   Rural decline

D   Rural growth

Localités - Administration

Souligné rouge: localités ou sites sélectionnés dans les
    Guides Michelin "Hôtels et Restaurants"

Cadre rouge : plans de villes traités dans ces
    mêmes guides

(▲) localités ou sites sélectionnés dans le
    Guide Michelin "Camping-Caravaning"

(25) : altitude de la localité

Limites : départements, provinces, cantons suisses. Frontière, douane

St Jean (25 ▲)

Ⓟ ... Préfecture
ⓢⓟ ... S. Préfecture
ⓒ ... Chef-lieu de Canton

Signes conventionnels
Repères

Hôtel ou restaurant isolé .... ⊡	Tour ou pylône de télécommunications	**Pittoresque Curiosités**
Terrain de camping situé .... ▲		Parcours pittoresque ....
Puits de pétrole ou de gaz .. ▲	Église ou chapelle........	Table d'orientation ....
Carrière - Mine .... ᴗ ⚒	Cimetière - Calvaire	Panorama ....
Usine - Château d'eau..... ⚙	Château - Ruines	Point de vue ....
Fort - Phare.............. ✿ ♨	Mon.t commémoratif..... Mon.t	Château ....
Établissement hospitalier .... ⊞	Moulin à vent	Église ou chapelle......
Barrage...................	Maison forestière......... ▪ MF	Ruines ....
Téléphone de secours (borne d'appel isolée)	Forêt ou bois .... Forêt domaniale	Autre curiosité........... ▲

Urbanization
1. The movement of people from rural to urban areas over a period of time.
2. The growth in size of population and area of cities as a result of this movement of people.

Counter-urbanization
1. The movement of people out of the big cities over a period of time.
2. The growth of population and area of settlements in rural areas.

Urban growth/sprawl
The growth of individual towns and cities.

Rural growth
The growth of population in rural areas.

Urban renewal
The redevelopment of areas of the city that have fallen into decay.

Metropolitan area
The built-up area of major cities; central zone (CBD), inner city and outer suburbs.

B Key words and definitions ▲

E Urban decline

F Urban growth/sprawl

Urbanization (2)

Millionaire cities of the eighties

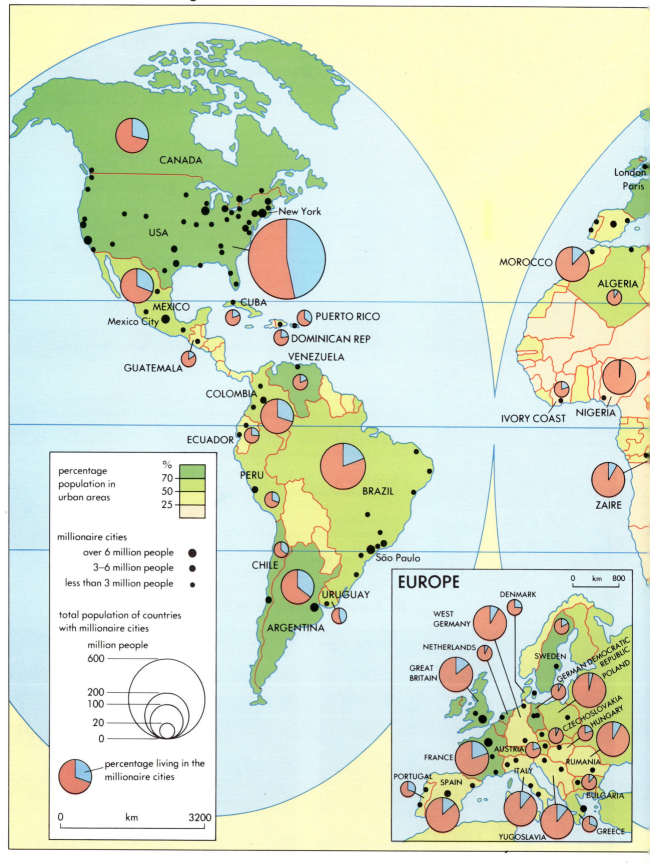

14

percentage population in urban areas

%	
70	
50	
25	

millionaire cities

- over 6 million people ●
- 3–6 million people •
- less than 3 million people ·

total population of countries with millionaire cities

million people

600
200
100
20
0

percentage living in the millionaire cities

0   km   3200

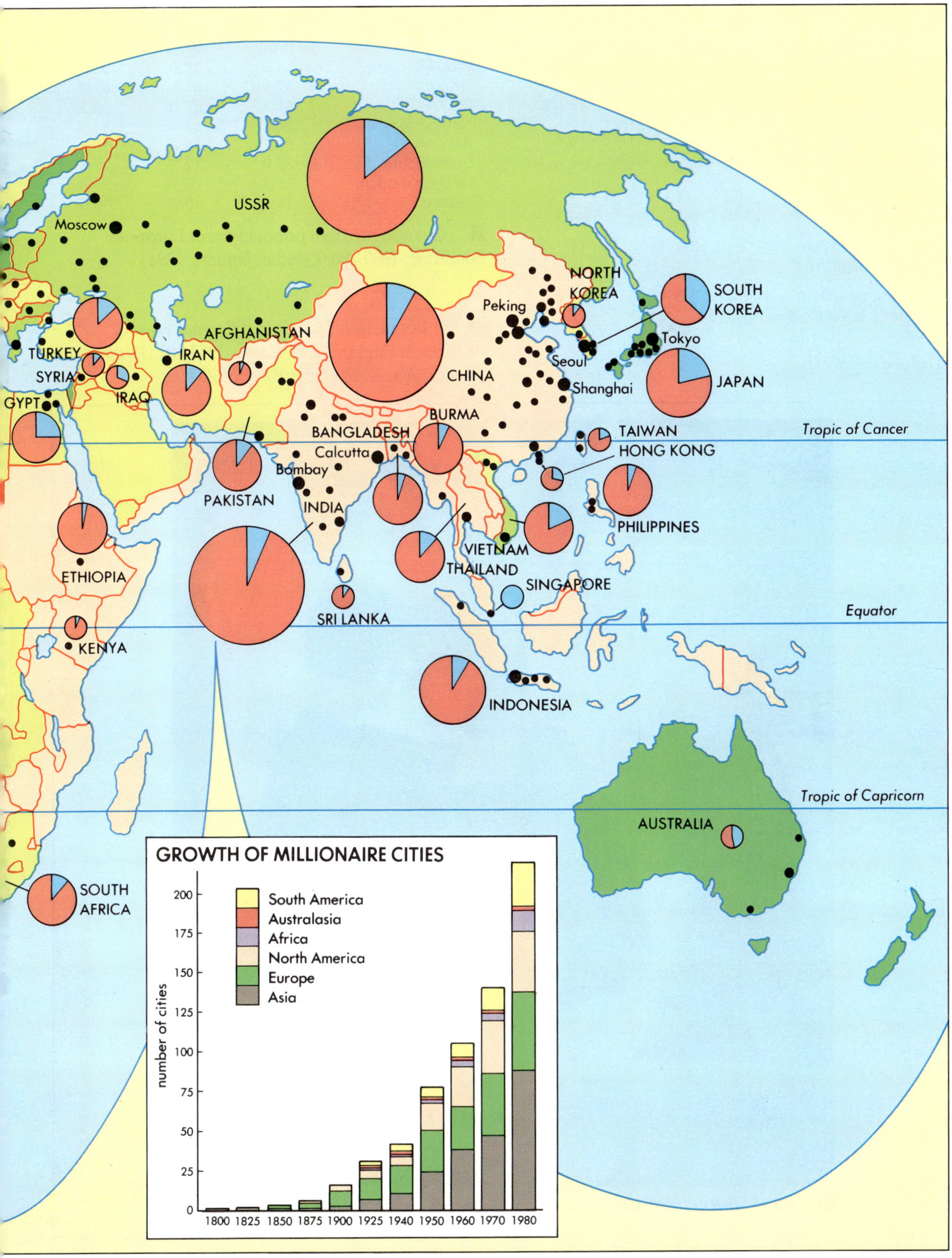
USSR
Moscow
TURKEY
SYRIA
IRAQ
IRAN
AFGHANISTAN
GYPT
PAKISTAN
INDIA
Bombay
Calcutta
BANGLADESH
BURMA
CHINA
Peking
Shanghai
Seoul
NORTH KOREA
SOUTH KOREA
Tokyo
JAPAN
TAIWAN
HONG KONG
PHILIPPINES
VIETNAM
THAILAND
SINGAPORE
SRI LANKA
INDONESIA
ETHIOPIA
KENYA
SOUTH AFRICA
AUSTRALIA
Tropic of Cancer
Equator
Tropic of Capricorn
GROWTH OF MILLIONAIRE CITIES
South America
Australasia
Africa
North America
Europe
Asia
number of cities
0
25
50
75
100
125
150
175
200
1800 1825 1850 1875 1900 1925 1940 1950 1960 1970 1980

G Millionaire cities of the eighties (*Based on 'Millionaire cities of the seventies' with revised data, Geographical Magazine, May 1978*)

Leaving Europe's big cities

Type of district	Population 1981 (millions)	Increase or decrease % (1971–81)
Inner London boroughs	2.5	−18.4
Outer London boroughs	4.2	−6.4
Large cities (over 175 000 in 1971)	2.8	−5.9
New Town districts	2.2	+9.7
Resorts and seaside retirement districts	3.3	+11.1
Industrial districts in Midlands, E. Anglia and South	3.3	+1.6
England and Wales	49.0	−0.6

Country	Urban population as a % of total population		
	1960	1970	1980
West Germany	54.3	52.0	53.1
France	57.9	61.1	62.5
Italy	61.1	65.0	65.2
Netherlands	66.7	65.4	64.0
Belgium	56.7	56.4	55.7
Luxembourg	22.9	22.5	21.4
United Kingdom	78.7	78.1	77.8
Ireland	35.6	40.1	50.0

H Change in urban populations of European countries, 1960–80 (*Geofile, January 1984*)

◀ **I** Population growth in different types of districts in England and Wales, 1971–81 (*Population Trends, No. 27, Spring 1982*, HMSO)

The most famous drink in the world is in the fastest growing city in the country.

MILTON KEYNES

J Advert for Milton Keynes – a new town (*Milton Keynes Development Corporation*)

K Percentage of Denmark's population living in urban areas, 1960–80 (*Court, 1985*)

	1960	1970	1975	1980
Capital	27.5	27.9	25.5	24.9
Other urban				
19 999 +	21.0	23.2	23.2	22.7
10 000–19 999	6.2	6.2	6.9	5.8
5 000– 9 999	4.0	3.4	4.3	4.6
1 000– 4 999	7.2	10.6	13.7	14.4
200– 999	8.1	8.5	4.5*	4.5*
Outside urban areas	26.0	20.1	21.9	23.1

* Urban area sizes 500–999

L Areas of population increase and decrease in the Paris region (*Hall and Ogden, 1985*) ▼

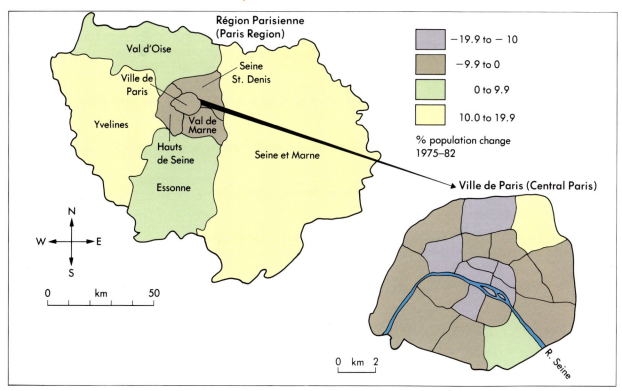

Région Parisienne (Paris Region)

Val d'Oise
Ville de Paris
Seine St. Denis
Yvelines
Val de Marne
Hauts de Seine
Seine et Marne
Essonne

Legend:
- −19.9 to −10
- −9.9 to 0
- 0 to 9.9
- 10.0 to 19.9

% population change 1975–82

Ville de Paris (Central Paris)

R. Seine

0 km 2

0 km 50

N W E S

I used to live in Streatham just south of central London until I got married. We went to live in Tadworth a beautiful place 5 km south of the Greater London boundary. Although I like living in the area, I hate to think of how much of my life has been wasted on British Rail. Perhaps if I were younger, I'd move back to the city.

According to population data the number of people living in some larger European cities is falling steadily.

M George Black, accountant, near retirement age

N News item: 'Towards 1992 . . .'

In the Green Belt (1)

The national scene

Purposes of Green Belts:
- to check the spread of further urban development
- to prevent neighbouring towns from merging into one another
- to preserve the special character of towns
- to safeguard the surrounding countryside from further encroachment
- to assist in urban regeneration

(1988 Government circular)

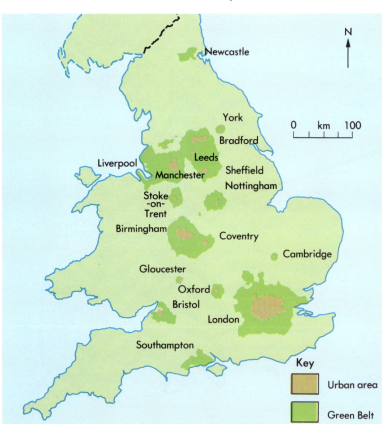

A Statutorily approved Green Belts in England and Wales, 1986 (*Munton, R, 'Green Belts – the end of an era', Geography, 1986*)

Circular 14/84 is a government pronouncement on Green Belts:

"The government continues to attach great importance to Green Belts, which have a broad and positive planning role in checking the unrealistic sprawl of built-up areas, safeguarding the surrounding countryside from further encroachment and assisting in urban regeneration. There must continue to be a general presumption against inappropriate development within Green Belts."

B Government circular 14/84 (*HMSO, 1984*)

C Metropolitan Green Belt, 1975 and 1986 (*Munton, R. 'Green Belts — the end of an era', Geography, 1986*)

Land use in the Metropolitan Green Belt

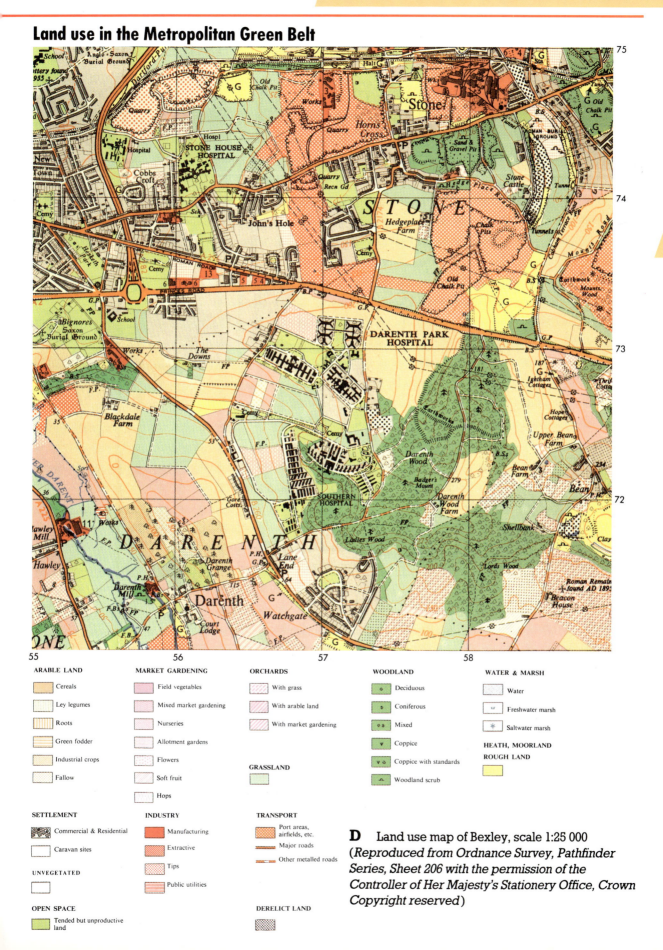

ARABLE LAND
- Cereals
- Ley legumes
- Roots
- Green fodder
- Industrial crops
- Fallow

MARKET GARDENING
- Field vegetables
- Mixed market gardening
- Nurseries
- Allotment gardens
- Flowers
- Soft fruit
- Hops

ORCHARDS
- With grass
- With arable land
- With market gardening

GRASSLAND

WOODLAND
- Deciduous
- Coniferous
- Mixed
- Coppice
- Coppice with standards
- Woodland scrub

WATER & MARSH
- Water
- Freshwater marsh
- Saltwater marsh

HEATH, MOORLAND ROUGH LAND

SETTLEMENT
- Commercial & Residential
- Caravan sites

UNVEGETATED

OPEN SPACE
- Tended but unproductive land

INDUSTRY
- Manufacturing
- Extractive
- Tips
- Public utilities

DERELICT LAND

TRANSPORT
- Port areas, airfields, etc.
- Major roads
- Other metalled roads

D   Land use map of Bexley, scale 1:25 000 (*Reproduced from Ordnance Survey, Pathfinder Series, Sheet 206 with the permission of the Controller of Her Majesty's Stationery Office, Crown Copyright reserved*)

The impact of the M25

Green Belt plan for new racetrack

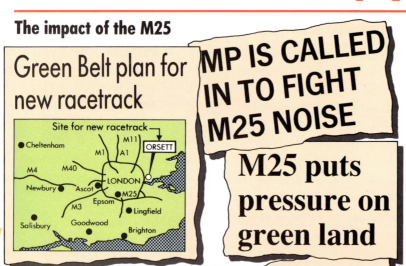

E   The impact of the M25

MP IS CALLED IN TO FIGHT M25 NOISE

M25 puts pressure on green land

Homes for farm split by the M25

Ring that would not fit

The M25 London orbital motorway, already so congested that parts of it are being widened, encircles nearly one-eighth of the UK population.

MOTORWAY PLAN COULD BOOST JOBS

MOTORWAY IMPACT WORRIES RESIDENTS

'Rape of the countryside'

F   A junction of the M25

Badgers lose out!

. . . or do they?

The M25 is almost entirely contained within the green belt and has the effect of altering the visual impression made by a reasonably intact girdle of open countryside as well as fragmenting the belt so that it might become more susceptible to development and lack of planning protection. The debate about the green belt will be fast and furious as planners defend their patch and developers work hard to find reasons why development should take place. The role of the Secretary of State for the Environment will be crucial and there are real fears that the green belt cannot withstand the windfall gain prizes that development will represent.

G   Extract from *M25: The Mighty Motorway* by J Whitelegg Winter 1986 no. 140

M25: an aid to tourism

The M25 orbital road is generating new trips as well as diverting those which would have been made by other routes. The importance of this for tourism should not be underestimated. The M25 is already helping to spread tourist traffic to attractions which previously suffered from poor accessibility. The English Tourist Board regards the M25 as the most valuable road built in Britain for more than a decade. The M25 assists the tourist board by giving the tourist – whether arriving by air at Heathrow or Gatwick or by sea at the Channel ports, whether by coach or car – the option of avoiding London and easily joining one of the radial routes to north, south, west or east. The distribution of tourists to places outside London to such 'honeypots' as Stratford-upon-Avon and the Lake District, spreads the economic benefits of tourism throughout the country.

Regionally the M25 is a valuable tourist asset. The map below shows a selection of tourist destinations which may often be more easily reached by the M25 than by driving through London. They include a varied range of attractions such as country parks, gardens, sporting venues, historic buildings and museums.

H Extract from *The M25 Orbital Motorway*, Department of Transport, 1986

Motorway	M25
M25 junction numbers are printed in blue, other motorway junction numbers are printed in black	3 / 1
Major routes in London	A3
British Rail stations with park-and-ride facilities	⇌
Motorway Service Areas	♦
Places of interest open to the public	15*

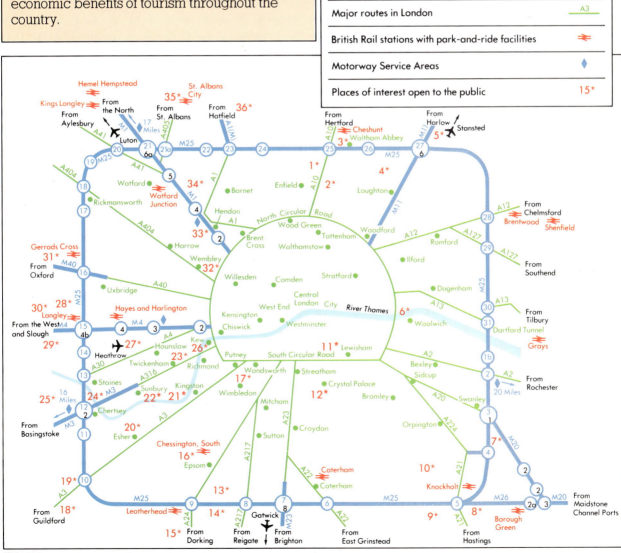

I Map showing tourist destinations easily reached by the M25 (*Department of Transport, 1986*)

The Golden Triangle

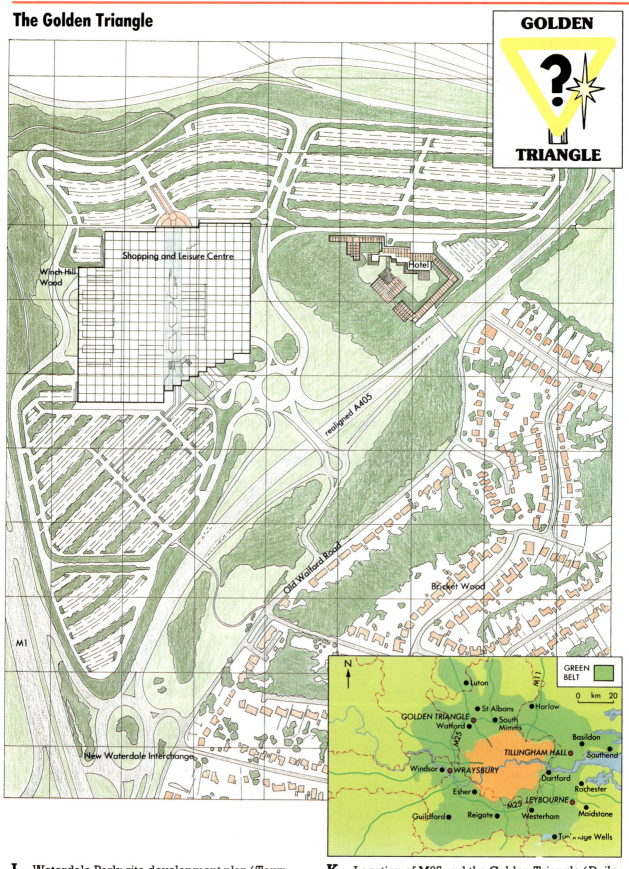

J Waterdale Park: site development plan (*Town and City Properties (Development) Ltd, 1986*)

K Location of M25 and the Golden Triangle (*Daily Telegraph, 16 June 1989*)

Waterdale Park: the proposed development

- A retail and leisure complex of regional importance.
- Main elements will be – shopping, leisure facilities, a hotel and car parking.
- The location will enable people to come from a large catchment area.
- New substantial tree planting will extend existing woodland and reduce traffic noise and unsightly views of the M1 and A405 from Bricket Wood.
- The development will provide about 750 000 square feet of shopping in a single building.
- There will be three major department stores, four other large stores and 30 small shops selling specialist merchandise.
- About 5500 car parking spaces will be provided in two main areas.
- The development is also designed for access by public transport – bus lay-bys, shelters and turn-round facilities.
- The site is fully accessible to the disabled.
- The building will be kept low, minimizing the visual impact from Bricket Wood and the surrounding countryside.
- There will be a food court with seating for 500 people.
- There will be a supervised children's recreational area.
- There will be a multi-screen cinema.
- Play areas for use in fine weather will be provided.
- The roof will be designed with artificial grass as a sport and recreational area.
- Planting of trees around the building will soften the appearance and the general effect will be of a light, low-scale development set in woodland.
- All servicing to the development will be by separate service road.

Triangle is Green Belt 'exception'

By John Grigsby
Local Government
Correspondent

THE UNIQUE setting of the "Golden Triangle" at the junction of two major motorways near St Albans is sufficiently exceptional to overcome the general presumption against development in the Green Belt, a public inquiry was told yesterday.

The progress of the inquiry, into a proposal to build a giant out-of-town shopping and leisure centre, is being watched closely by conservation groups and developers who are anxious to see whether there are any circumstances in which the Government will drop its general opposition to such shopping centres in the Green Belt.

Nearly all the sites around the M25 proposed by developers for similar centres lie within it.

Town and City Properties (Developments) wants to build a 1,100,000 sq ft complex, including more than 670,000 sq ft of shops, a 250-bed hotel, leisure facilities including a 10-screen cinema, and parking for 5,500 cars on the 124 acre site bordered by the M25 and M1 and the A405.

Mr Michael Harrison, QC, for Town and City, said: "No other site in the country has these attributes, offering unparalleled accessibility to over three million people within 30 minutes driving time."

L *Daily Telegraph,*
3 February 1988 ▲

Villagers voice their protest

PROTESTORS were out in force at the start of the Golden Triangle inquiry.

Armed with banners the demonstrators — members of SOCEM (Save Our Society from Environmental Mess) — were mostly residents of Bricket Wood.

They are arguing that Bricket Wood will change from a pleasant village to an urban sprawl if the development goes ahead.

SOCEM has been busy distributing leaflets and plans to have members sitting in the inquiry every day wearing badges as a permanent demonstration of their opposition.

M *Watford Observer,*
5 February 1988

The new country town plan

A consortium of housing contractors is planning to build a series of 12 small new towns or large villages on green field sites within easy commuting distance of London. They have already identified "target" locations which are being kept secret, although some of them are known to be in the green belt.

Their plan is to build entirely new and self-contained villages, well away from any existing population centres, and to build them complete with roads, shops, pubs, churches and schools, with little or no help from the local authorities.

The villages will have populations of 15,000 to 20,000 and the consortium has tried to pick sites which are neither prime agricultural land nor areas of outstanding natural beauty, but surplus land such as disused airfields, all within a 30-mile radius of London.

A The plan (*Guardian*, 22 July 1983)

B One view of urban development (*Private Eye*)

Housing requirements in the south-east 1981–91

720 000 — to accommodate the increase in households. (Dept of Environment estimate)

36 000 — to allow for mobility / to combat overall shortage in London (GLC figure)

68 000 — to replace homes in London which have reached the end of useful life (GLC figure)

60 000

Total 884 000 homes

Housing completions in the south-east 1981–91

600 000

SERPLAN forecast

Source: Consortium Developments, press release, 8 May 1985

C Why? The developer's view

George Nicholson, Chairman of GLC Planning Committee
'We are convinced our figures show there is no need for this development.

'There is enough land in London to build the houses that need to be built to the end of the century.'
Guardian, 14 May 1985

Forecast household growth in the south-east, 1981–91

540 000

SERPLAN forecast, based on County Council plans
Source: Consortium Developments, press release, 8 May 1985

'. . . the proposal to construct more housing in the green belt and to green over derelict land in inner London would strengthen the reality of "two nations." Without public investment, including housing development, in our inner cities the countryside will indeed be for the rich and the city left for the poor.'
Guardian, 24 August 1985

Chris Shepley, Deputy Planning Officer, Greater Manchester Council
'Anybody can have a glamorous and profitable time building trendy new villages in the leafy Essex green belt, but so far as your average long-term unemployed, increasingly bitter and resentful northern inner city teenager is concerned, they may as well be on the moon.'
Guardian, 14 May 1985

D Why not? the planner's view

A new country town at Tillingham Hall

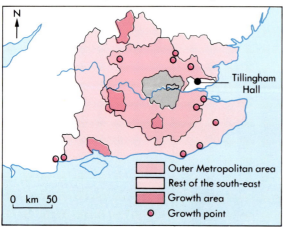

Tillingham Hall

- Outer Metropolitan area
- Rest of the south-east
- Growth area
- Growth point

E Growth areas in south-east England (*Consortium Developments Ltd, Tillingham Hall Outline Plan, May 1985*)

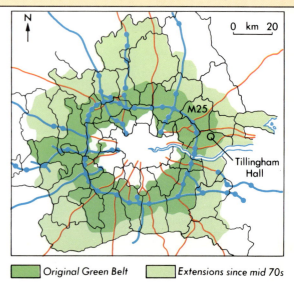

M25

Tillingham Hall

Original Green Belt   Extensions since mid 70s

F The location of Tillingham Hall (*Consortium Developments Ltd, Tillingham Hall Outline Plan, May 1985*)

G Tillingham Hall farm

The site for the new country town is some 761 acres in extent, immediately to the south of West Horndon. It is 5 miles east of the built-up edge of Greater London and 4 miles east of the M25 interchange with the A127 (London-Southend) trunk road. It is 5 miles west of Basildon new town.

The boundaries of the site are Dunnings Lane to the west, the London-Southend railway line to the north (beyond which lies West Horndon), the A128 to the east and part of the Mar Dyke to the south. To the south east of the site is the village of Bulphan.

The use of the land is currently medium grade agricultural land. Use of this agricultural land for development will have minimal adverse effect on surrounding farm holdings. It is proposed to define the boundary of the town to prevent further growth and to landscape the town. It is also proposed to devote a major part of the site to public parkland and water features. The site does not lie in an Area of Outstanding Natural Beauty.

I Proposed location of Tillingham Hall (*Reproduced from Ordnance Survey, 1:250 000 Routemaster Series, Sheet 9 with the permission of the Controller of Her Majesty's Stationery Office, Crown Copyright reserved*) ▼

H The Tillingham Hall site (*Consortium Developments Ltd, Tillingham Hall Outline Plan, May 1985*)

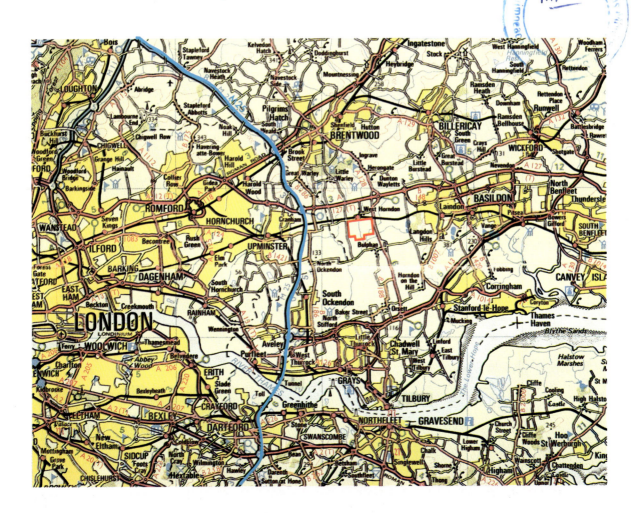

New country towns (2)

Tillingham Hall: the plan

The Tillingham Hall plan is for a country town with 5000 homes, as well as lakes, employment areas providing 2000 jobs, a high street, with shops, small offices and workshops, extensive new landscaping, schools, community centres and health facilities (*Consortium Developments press release, 7 May 1985*)

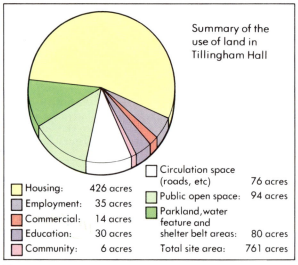

Summary of the use of land in Tillingham Hall

☐ Housing:	426 acres
☐ Employment:	35 acres
☐ Commercial:	14 acres
☐ Education:	30 acres
☐ Community:	6 acres

☐ Circulation space (roads, etc)	76 acres
☐ Public open space:	94 acres
☐ Parkland, water feature and shelter belt areas:	80 acres
Total site area:	761 acres

J   Summary of the use of land in Tillingham Hall

L   Employment and housing

☐ Housing		☐ Commercial	
☐ Employment		☐ Parkland	
☐ Education and community		☐ Water	
☐ Woodland Shelter belt		☐ Parking	

Based on the Ordnan... permission of the Co... Office. Crown Copyr...

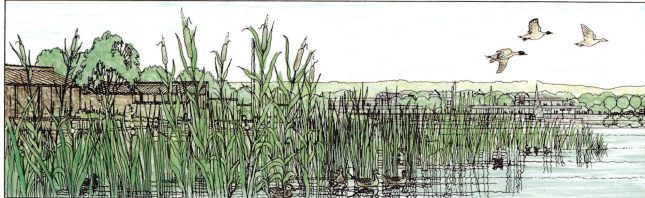

M   View of the centre of the country town, looking north

K   Planned layout of Tillingham Hall

10000 map with the
r Majesty's Stationery
d.

N   Water-related housing

O   High Street

P   Viewing to countryside

Source: Consortium Developments Ltd, *Tillingham Hall Outline Plan*, May 1985

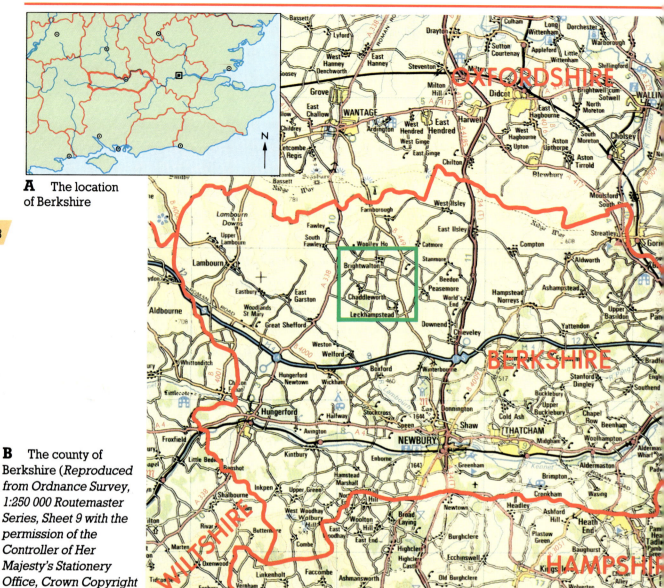

A The location of Berkshire

28

B The county of Berkshire (*Reproduced from Ordnance Survey, 1:250 000 Routemaster Series, Sheet 9 with the permission of the Controller of Her Majesty's Stationery Office, Crown Copyright reserved*)

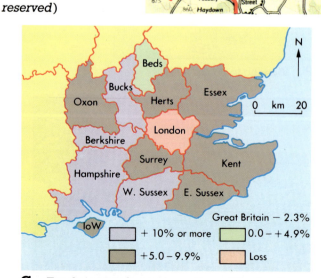

C Employment change in south-east England, 1971–81 (*Source of data: SERPLAN*)

Great Britain – 2.3%

+ 10% or more   0.0 – +4.9%
+5.0 – 9.9%   Loss

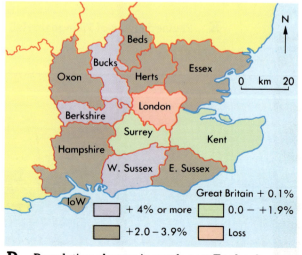

D Population change in south-east England, 1978–83 (*Source of data: Central Statistical Office*)

Great Britain + 0.1%

+ 4% or more   0.0 – +1.9%
+2.0 – 3.9%   Loss

E A 'rural' landscape, 1986

Beyond the Green Belt (2)

More new housing for Central Berkshire?

A   Woosehill, Wokingham before housing development, 1976

B   Woosehill, Wokingham after housing development, 1986

The County Council view

"POLICY H1   The County Council proposes that there should be further growth in the housing stock and provision will be made for the development of land for approximately 31 700 dwellings.

POLICY H4   No further land for housing on major sites will be released before 1983."

From the *Central Berkshire Structure Plan*, submitted to the Secretary of State for the Environment for approval, June 1978

C   How much new housing should Central Berkshire have?

The Department of the Environment view

"The Secretary of State considers that the strategic needs of the South East region require provision to be made in Central Berkshire to accommodate pressures arising elsewhere in the region outside London, and to complement the restraint policies applied in Green Belt areas such as East Berkshire. He has therefore modified Policy H4 to indicate that planned provision should be made sufficient for the construction of about an additional 8000 dwellings from the mid-1980s."

From a letter from the Departments of the Environment and Transport to Berkshire County Council, 14 April 1980, explaining amendments made to the *Central Berkshire Structure Plan*

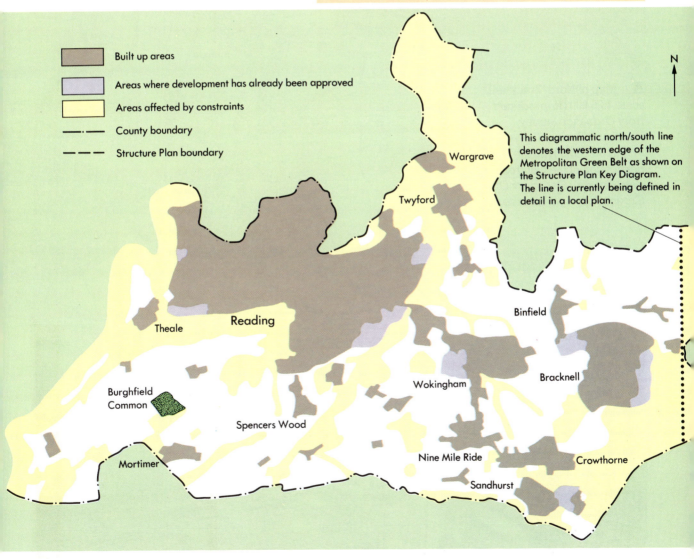

D   Development constraints in Central Berkshire (*Berkshire County Council, Berkshire Structure Plans, Central Berkshire, Land for 8000 houses: The choices, April 1982*)

Beyond the Green Belt (3)

Housing development in North Bracknell

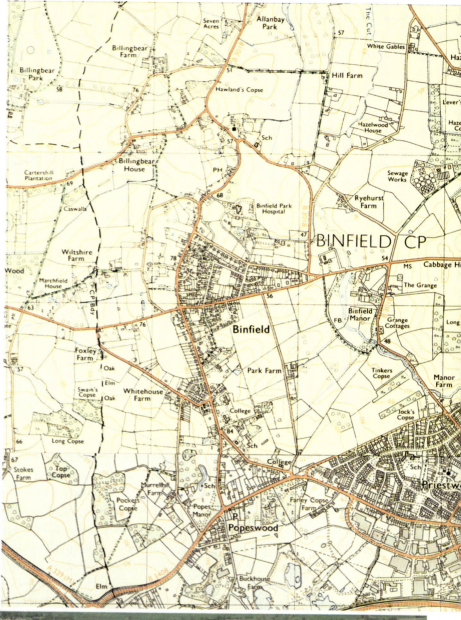

A   Map of North Bracknell, scale 1:25 000 (*Reproduced from Ordnance Survey, Pathfinder Series, Sheets SU 87/97 and SU 86/96 with the permission of the Controller of Her Majesty's Stationery Office, Crown Copyright reserved*)

B   North West Bracknell

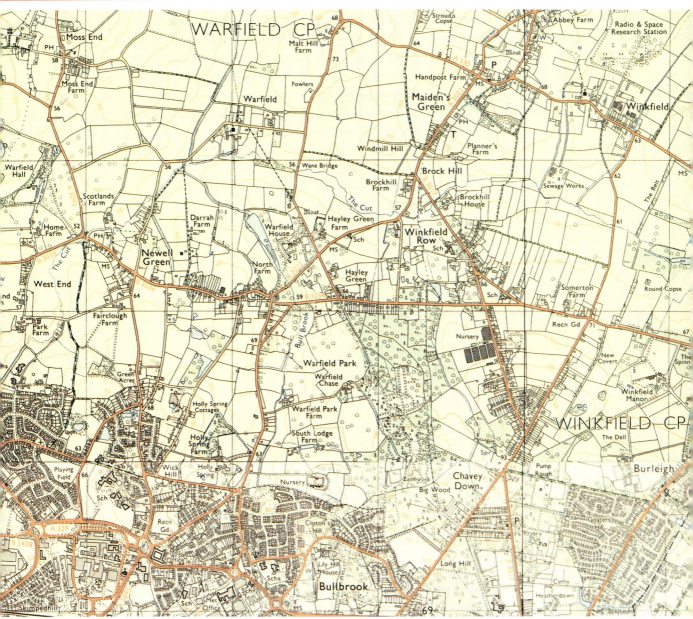

C North East Bracknell

Beyond the Green Belt (4)

Where should the new houses go?

In 1980, the Secretary of State for the Environment instructed Berkshire County Council to find land for 8000 more houses in the Central Berkshire area. The County Council did a study and in 1983 allocated 4000 of the 8000 houses to the North East Bracknell area. Bracknell District Council then started preparing a Local Plan to find sites for these houses and last year published a first consultation Plan for public comment.

Berkshire County Council has recently been looking at the North Bracknell housing allocation again when preparing the County Structure Plan Review. This structure plan looks at how much new development the County should take in the next 10 years. In this plan the County Council has taken account of the strong opposition to new houses in the North Bracknell area and reduced the figure from 4000 to 1500.

The options

The development of 1500 houses at North Bracknell will obviously have a big impact on the area. There is, however, some choice over where the housing could be built. To help show this, four options have been prepared, each one showing a different way of providing land for 1500 houses.

These options cannot cover all possibilities but show a range of realistic alternatives. The number of houses in each of the development areas is shown on the map.

A   Background to the 1500 houses

OPTION 1: NORTH OF BULLBROOK

By putting development in one area the countryside gap between Binfield and Bracknell is unaffected and better quality agricultural land is avoided. Additional traffic is concentrated in the Jigs Lane/Warfield Road area.

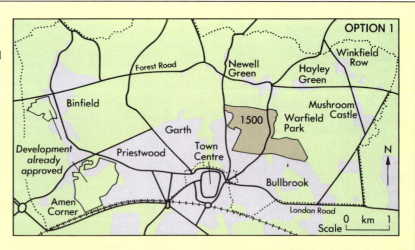

B   Option 1

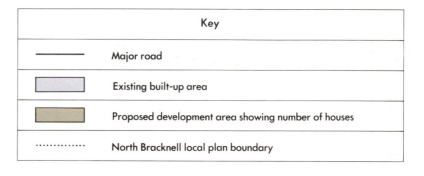

Key	
————	Major road
� box	Existing built-up area
▱	Proposed development area showing number of houses
·············	North Bracknell local plan boundary

OPTION 2: NORTH OF BULLBROOK AND NORTH OF PRIESTWOOD

This option divides new development and additional traffic between two areas. The Binfield/Bracknell gap is reduced and additional development pressure may be placed upon land north of Garth.

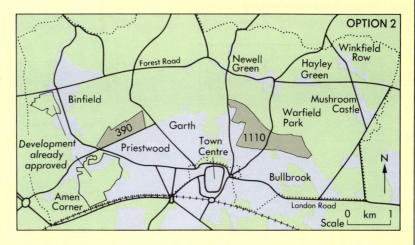

C Option 2

OPTION 3: NORTH OF BULLBROOK, NORTH OF PRIESTWOOD AND CARNATION NURSERY

This option develops land at Carnation Nursery allowing for the removal of the derelict glasshouses. It allows a more even dispersal of traffic, although additional development pressure may be placed on the land north of Garth.

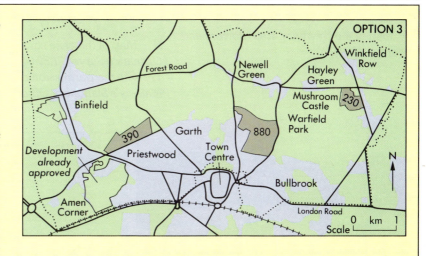

D Option 3

OPTION 4: NORTH OF BULLBROOK, NORTH OF PRIESTWOOD AND CARNATION NURSERY (ALTERNATIVE DISTRIBUTION)

This development pattern maintains the widest countryside gap between Bracknell and the Forest Road settlements. The Binfield/Bracknell gap is greatly reduced. At Carnation Nursery, new housing is limited to the derelict glasshouses but development pressure may be placed on adjoining land.

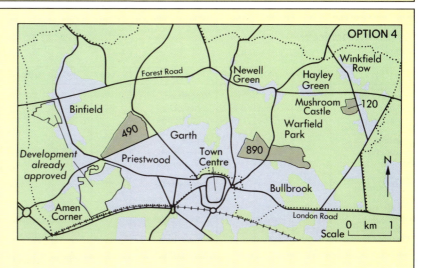

E Option 4

Source: Bracknell District Council Planning Department, North Bracknell Local Plan, *1500 Houses – The Options. Your Chance to Comment,* September 1986

The changing rural landscape

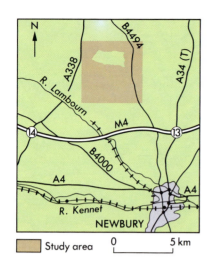

A Map of the Newbury area (*Bowers and Cheshire, 1983*)

B Map of the case study area (*Reproduced from Ordnance Survey, 1:50 000 Landranger Series, Sheet 174 with the permission of the Controller of Her Majesty's Stationery Office, Crown Copyright reserved*)

C Farm machinery

D Number of farmworkers (*Sources of data: Bowers & Cheshire/MAFF*)

	1947	1976	1981		1947	1976	1981
Hedges total				**Woodland**			
km	24.8	16.6	15.2	hectares	30.3	30.3	30.3
% change		−33.1	−8.4	% change		−	−
Visible footpaths/ tracks				**Arable**			
km	5.8	3.5	3.1	hectares	146.7	213.9	some
% change		−39.7	−11.4	% change		+45.8	decline
Ponds: number	8	1	1	**Permanent pasture**			
% change		−87.5	−	hectares	85.3	5.9	some
				% change		−93.1	increase
Working farms: number	4	2	2	**Ley pasture**			
% change		−50.0	−	hectares	24.8	30.0	some
				% change		+21.0	increase
New agricultural buildings		6	0	**Residential area**			
				hectares	7.9	13.4	14.0
				% change		+69.6	+4.5

F Blocked footpaths

E Countryside changes in the area, 1947–81 (*Bowers and Cheshire, 1983*)

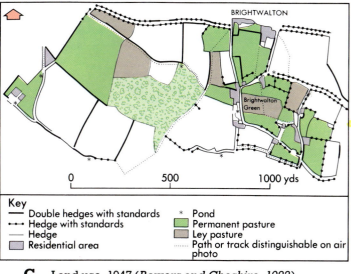

G Land use, 1947 (*Bowers and Cheshire, 1983*)

Key
— Double hedges with standards
•••• Hedge with standards
— Hedge
▨ Residential area
* Pond
▨ Permanent pasture
▨ Ley pasture
⋯⋯ Path or track distinguishable on air photo

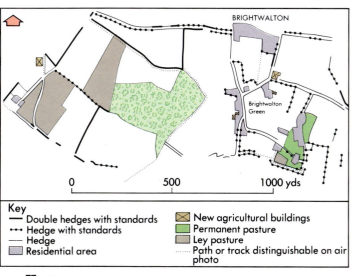

H Land use, 1976 (*Bowers and Cheshire, 1983*)

Key
— Double hedges with standards
•••• Hedge with standards
— Hedge
▨ Residential area
⊠ New agricultural buildings
▨ Permanent pasture
▨ Ley pasture
⋯⋯ Path or track distinguishable on air photo

I The rural landscape, September 1963 (scale 1:10 000)

J The rural landscape, September 1986 (scale 1:10 000)

K A farmer's view

The government and the EEC decided that they wanted the Common Market to be self-sufficient in food and they've paid us subsidies to increase our production. Our yields of wheat and barley have very nearly doubled in the last ten years. Today, we run through a field probably nine times - we put on two different herbicides, three or four different fungicides, two insecticides and three dollops of nitrogen fertilizer. We spend a fortune on a crop, in terms of chemicals, fertilizer, time and machinery. This is only worthwhile if the price is high. We've invested a vast amount of money, some of which has come from the taxpayer and we have accepted with gratitude what the taxpayer has offered: we've invested it in big machines, in modern equipment, modern buildings... and boosted production and efficiency enormously.

The changing village – Brightwalton Green

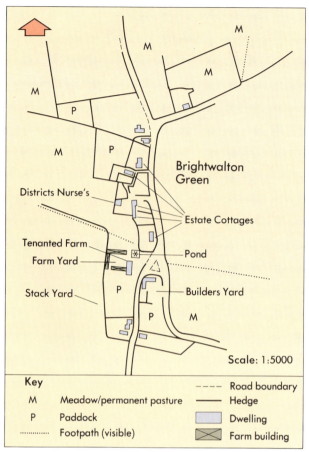

Brightwalton
Green

Districts Nurse's

Estate Cottages

Tenanted Farm

Farm Yard

Pond

Stack Yard

Builders Yard

Scale: 1:5000

Key

M	Meadow/permanent pasture	- - - -	Road boundary
P	Paddock	———	Hedge
··········	Footpath (visible)		Dwelling
		⊠	Farm building

A Brightwalton Green, 1947 (*Bowers and Cheshire, 1983*)

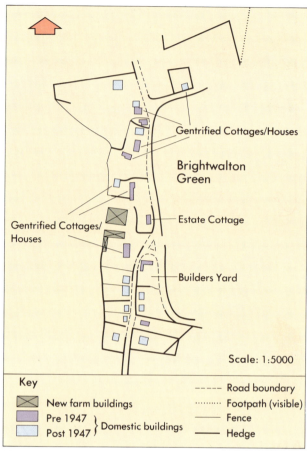

Gentrified Cottages/Houses

Brightwalton
Green

Gentrified Cottages/
Houses

Estate Cottage

Builders Yard

Scale: 1:5000

Key

⊠	New farm buildings	- - - -	Road boundary
	Pre 1947 } Domestic buildings	··········	Footpath (visible)
	Post 1947	———	Fence
		———	Hedge

B Brightwalton Green, 1983 (*Bowers and Cheshire, 1983*)

C Old Brightwalton

D New Brightwalton

38

The modern buildings we obviously use, the old buildings we try to get planning permission for and sell them off because we just don't need them and rather than let them decay, we either convert them or sell them to a builder for conversion. I think the converted buildings look rather nice, whereas before they were just falling down and an eyesore.

A local farmer

A family who have recently moved into the village

I'm still working; I'm in business. I do quite a lot of work in America and London Airport is very convenient from here - about 45 minutes in the car. The motorway is only 5-6 miles away.

We had to have about half an acre of garden; we'd have liked a paddock but we didn't get one. We wanted a cottage-type of house, in the country, with five bedrooms and two bathrooms.

It's just a fact of life. If it wasn't for people like me and the young professional people who want to be in the country, then houses of this sort would fall down and wouldn't be repaired and looked after properly. I think generally we all get on very well together.

They don't enter into village life like they should do. I think if you come into the country, you should enter into village life. But they've all got cars and they can all go shopping where they like, but the likes of us, we've got to stay put. We had a doctor, who lived in the village, and a nurse. We haven't a shop here now. You see, it's gradually gone. There aren't many farm workers now in Brightwalton. There's only one or two on each farm; they've got less and less.

An elderly woman who has lived in the village all her life

A local farmworker

The people who come into the village want to take over and the country people sit back and let them. They don't even want there to be council houses built in the village because this would spoil the village. They want to keep it their way. My sons wouldn't be able to afford to buy a house in this village because they've much too costly, so I think the country folk get less and less.

E Views on recent changes in the village (*Thames Television, Against the Grain, 1983*)

The Cairngorm environment

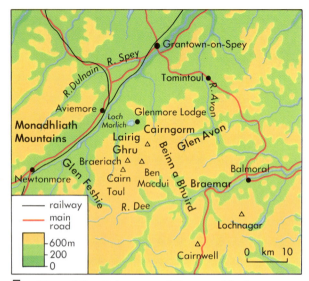

A Map of the Cairngorms (*Geographical Magazine, March 1980*)

The Cairngorms are unique in Britain:

- largest area of land above 1000 metres
- arctic–alpine climate of severe, cold winters and cool summers
- largest ski development and tourist complex
- rare birds, animals and plants.

B General view of the Cairngorm summits

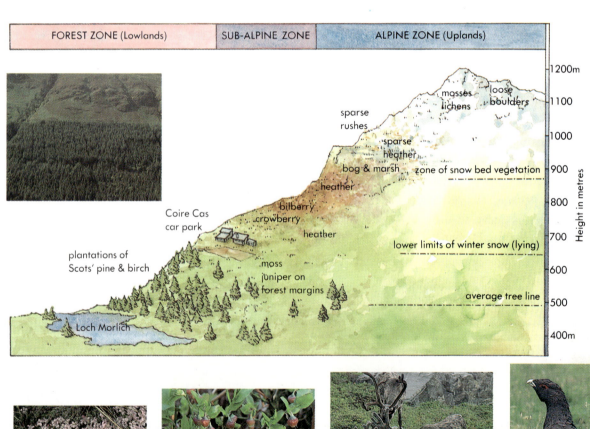

FOREST ZONE (Lowlands)	SUB-ALPINE ZONE	ALPINE ZONE (Uplands)

Labels on diagram: mosses, lichens, loose boulders — 1200m, 1100, 1000
sparse rushes
sparse heather
bog & marsh — zone of snow bed vegetation — 900
heather — 800
bilberry crowberry
Coire Cas car park
heather — 700
lower limits of winter snow (lying)
plantations of Scots' pine & birch
moss — 600
juniper on forest margins
average tree line — 500
Loch Morlich — 400m

Height in metres

C Altitudinal changes in vegetation and wildlife in the Cairngorms

D   Ptarmigan Restaurant near Cairn Gorm summit

E   Location of the Cairngorms

F   Tourist activities in the ▶
Cairngorms (*Spey Valley
Tourist Organization*)

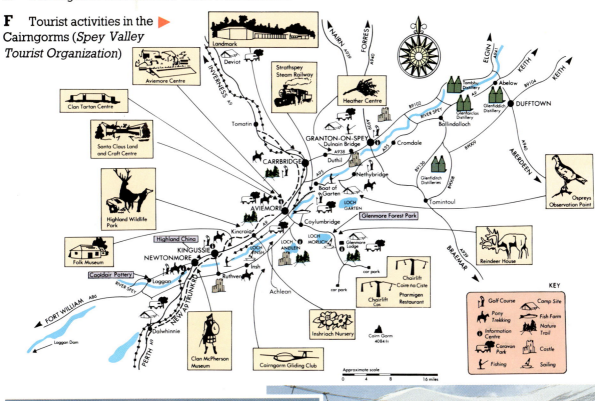

G   Summer hillwalkers

H   Winter skiers

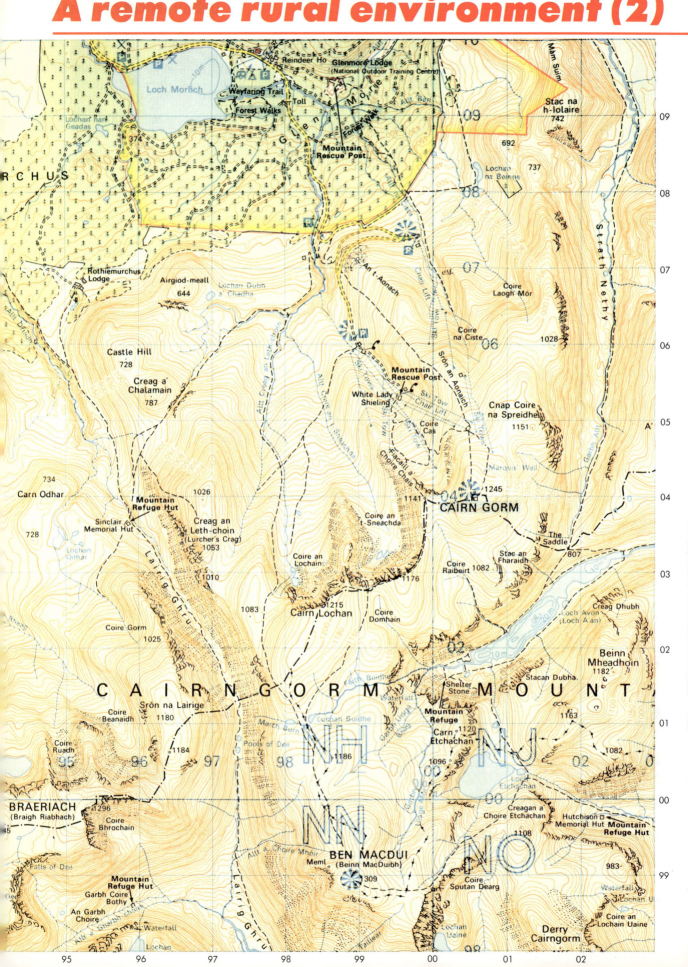

J   View of the northern corries – Coire an Sneachda and Coire an
Lochain, looking due east from a point above the Sinclair hut, 95 8036

◀ **I**   Extract from Ordnance Survey, 1:50 000 Landranger Series, Sheet
36 (Grantown and Cairngorm) (*Reproduced with the permission of the
Controller of Her Majesty's Stationery Office, Crown Copyright reserved*)

K   Footpath erosion on the summit path to Cairn Gorm

POLICIES FOR PROTECTION

'The main problem is that many tourists with conflicting requirements now threaten one another's enjoyment and the opportunities to enjoy the Cairngorms in these ways are therefore at risk . . .
Proposals for new developments add to problems. Some people propose new roads and chairlifts to provide more jobs and money, and to ease congestion on the ski slopes. However new access also creates problems. By using roads and lifts at Cairn Gorm, many walkers now go to places that most of them would not visit were it not for easy access.'

L   Policies for protection (*Watson, Geographical Magazine, December 1980*)

M   Coire Cas car park (989061) on a winter weekend

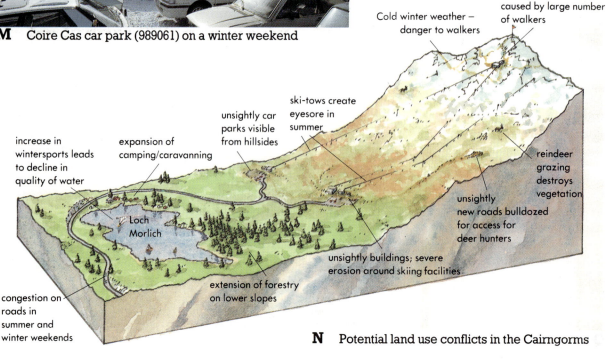

Cold winter weather – danger to walkers

erosion of footpath caused by large number of walkers

ski-tows create eyesore in summer

unsightly car parks visible from hillsides

expansion of camping/caravanning

increase in wintersports leads to decline in quality of water

reindeer grazing destroys vegetation

unsightly new roads bulldozed for access for deer hunters

unsightly buildings; severe erosion around skiing facilities

Loch Morlich

extension of forestry on lower slopes

congestion on roads in summer and winter weekends

N   Potential land use conflicts in the Cairngorms

O   Aerial view from Coire Cas car park (989061) to Cairn Gorm summit, looking SE

P   Water storage scheme

Q   Bulldozer smashing through pine forest

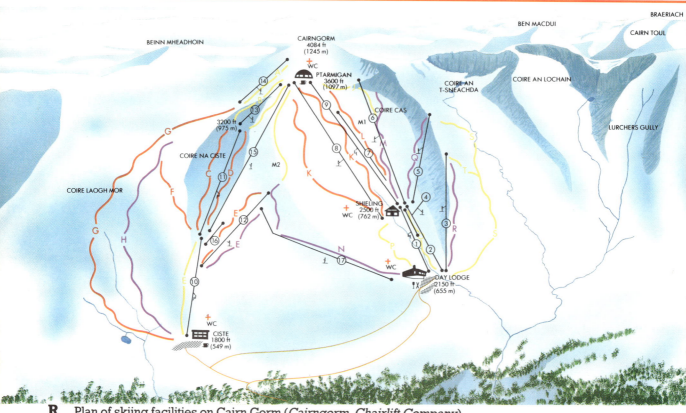

R Plan of skiing facilities on Cairn Gorm (*Cairngorm Chairlift Company*)

T Getting away from it all (*Richard Fisher*)

S A busy winter weekend

U Winter icing of weather recorder

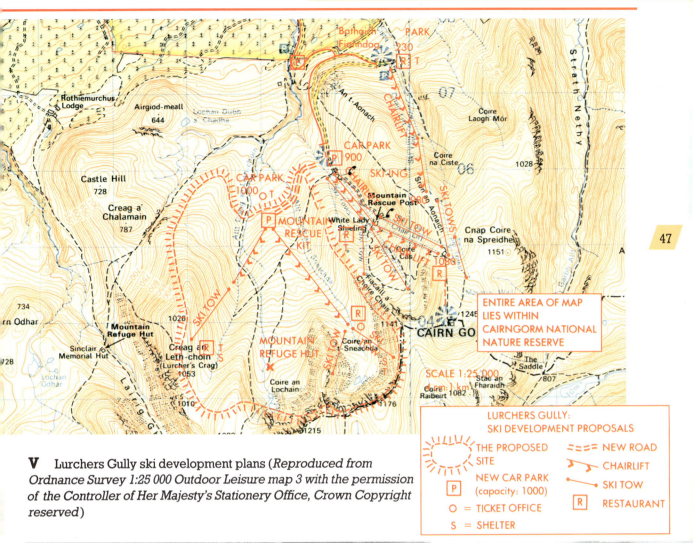

SCALE 1:25 000

LURCHERS GULLY:
SKI DEVELOPMENT PROPOSALS

THE PROPOSED SITE

NEW CAR PARK
(capacity: 1000)

P

O = TICKET OFFICE

S = SHELTER

NEW ROAD

CHAIRLIFT

SKI TOW

R   RESTAURANT

V Lurchers Gully ski development plans (*Reproduced from Ordnance Survey 1:25 000 Outdoor Leisure map 3 with the permission of the Controller of Her Majesty's Stationery Office, Crown Copyright reserved*)

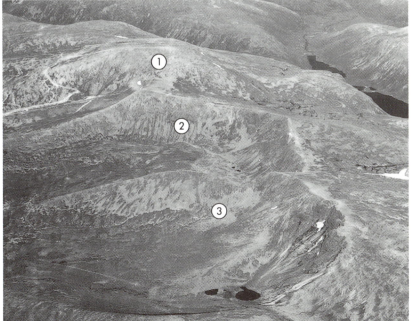

W Aerial view of northern corries looking east.   1 Cairn Gorm
2 Coire an Sneachda   3 Coire an Lochain

X Above Coire an Sneachda

Background to Turkey

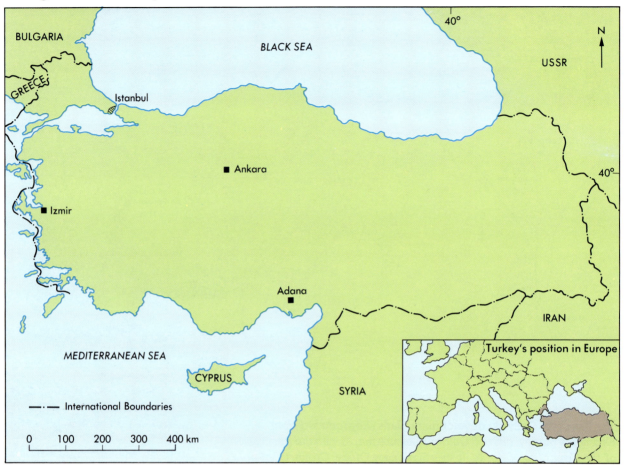

A   A Map of Turkey

Area: 781 000 sq km

Total population (1985):   51 428 514 (6 942 780 in 'European' Turkey and 44 485 734 in 'Asian' Turkey)

GNP per head (1985): $US1080

Share of wealth:
   poorest 20% of population share 3.5% of the wealth
   richest 20% of population share 56.5% of the wealth

Position of women: 35% of women work outside the home

Political prisoners: 15 000 in gaol
   (*Amnesty International, 1986*)

Religion: 98.99% of total population are Muslims

Education (1984):   38% of age group enrolled in secondary school (47% of boys and 28% of girls)

Life expectancy (1985):   people, on average, live for 64 years

Population per doctor (1980):   1630

Infant mortality rate (1985):   84 per 1000 live births

B   Facts about Turkey (*Sources of data: Whitaker's Almanac, 1987; World Development Report, OUP, 1987*)

Year	% urban population
1927	38.8
1935	38.5
1940	35.3
1945	40.3
1950	44.6
1955	45.7
1960	48.4
1965	50.7
1970	52.8
1980	53.3
1985	55.0

C   Urbanization in Turkey

A migrant is a person who crosses a boundary to work or be supported by someone who works. They do this for an extended period of time, not just a few months.

D   What is a migrant? (*Open University, 1983*)

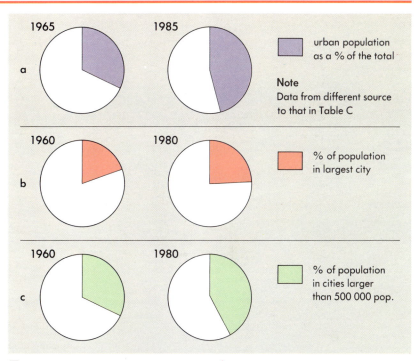

a — urban population as a % of the total (1965, 1985)

Note Data from different source to that in Table C

b — % of population in largest city (1960, 1980)

c — % of population in cities larger than 500 000 pop. (1960, 1980)

E   Changes in urban population in Turkey (*World Development Report, OUP, 1987*)

F   Ankara

G   Gecekondu housing in Ankara

H   Anatolia

I   Tourism in Turkey

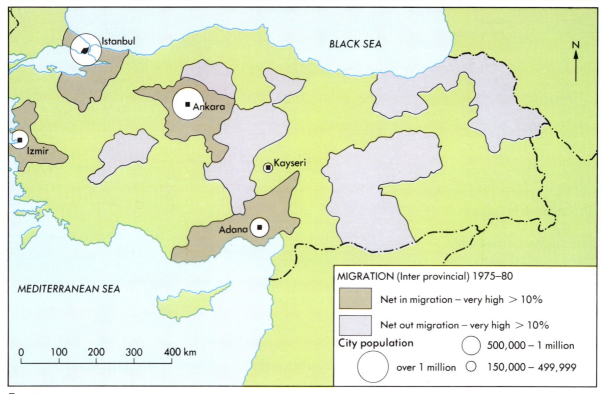

J   Movement of people (*Open University, 1983*)

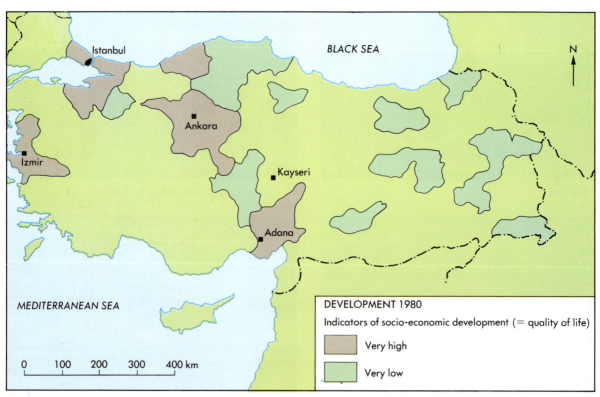

K   Contrasts in development (*Open University, 1983*)

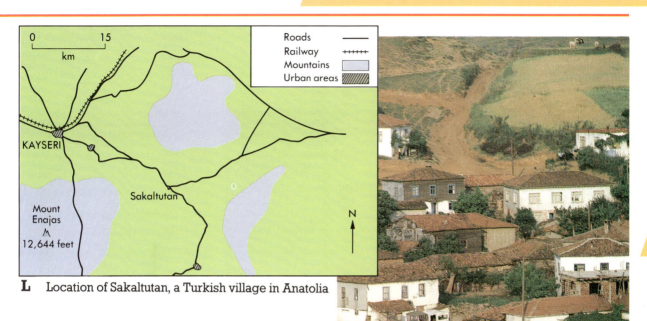

L Location of Sakaltutan, a Turkish village in Anatolia

Roads ————
Railway +++++++
Mountains
Urban areas

KAYSERI

Mount
Enajas
⅍
12,644 feet

Sakaltutan

N

0   15
km

M A view of Sakaltutan

> 40 years ago we worked
> only in our homes and
> villages.
> We didn't know other
> places, we couldn't even
> get shoes.

> Those who have gone to
> Ankara or Istanbul leave
> their land and their
> brothers or relatives farm
> it and send them food.
> They don't sell their
> land.

> Of old, 8 or 10 people would work
> together. We ploughed the land
> with oxen. It took a lot of man power.
> Now tractors have come along. A
> man can have his land ploughed
> for money. All he needs to do is sow
> and one worker is enough for that
> the rest go to cities or Europe to work
> it's much better.

> The road they built in 1952
> cut journey time from
> Sakaltutan to Kayseri
> from 10 hours to 20
> minutes.

> Over a hundred village men
> work outside the area. The
> first who went to work in
> the cities found things better
> there. They moved their
> homes. Others work in Turkish
> towns and send back their
> money. Some have their
> homes and their money in
> town.

> Now the young people are
> used to going away.
> They don't wear what
> we used to wear; they
> get quality clothes.

N Ali Osman – school teacher
in Sakaltutan village school for
over 30 years (*BBC, Sakaltutan –
A Time of Change*)

A new life in the city

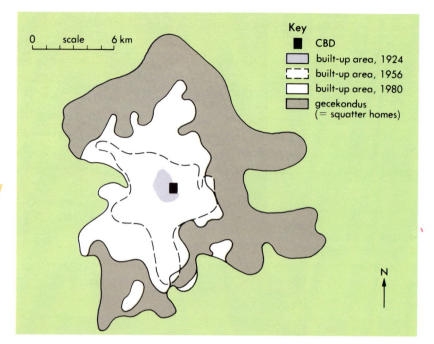

Year	Population
1923	20 000
1927	74 500
1935	123 000
1940	157 250
1950	289 250
1955	454 000
1960	650 000
1970	1 200 000
1975	1 700 000
1980	2 300 000
1985	3 000 000

P Growth of Ankara's population, 1923–85

Key
- ■ CBD
- built-up area, 1924
- built-up area, 1956
- built-up area, 1980
- gecekondus (= squatter homes)

O Growth of Ankara, 1924–80 (*Drakakis–Smith and Fisher, 1975*) ◄

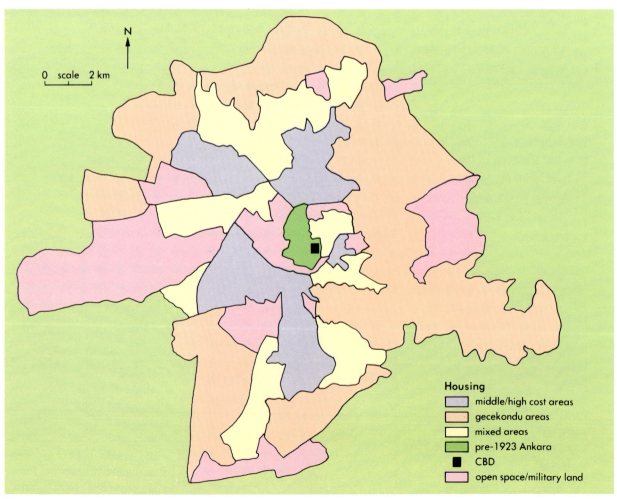

Housing
- middle/high cost areas
- gecekondu areas
- mixed areas
- pre-1923 Ankara
- ■ CBD
- open space/military land

Q Land use in Ankara, 1974 (*Drakakis–Smith and Fisher, 1975*)

'As in most countries, once the squatter hut is <u>erected</u>, eviction and <u>demolition</u> can involve long, costly <u>legal procedures</u>'

'A gecekondu literally means a dwelling 'built overnight.' In 1966 the Gecekondu Law passed by the Turkish government said a gecekondu is a 'dwelling without a licence on a piece of land for which the used does not have a title.'

(Payne, 1984)

'The Turkish <u>gecekondu</u> is a squatter house purpose–built on public or private land without legal authorization. It is not necessarily a slum.'

(Open University, 1983)

'Ankara municipality were given the right to demolish squatter housing by an Act in 1949. One particular squatter is known to have re-erected his house on the same site four times over.'

(Drakakis – Smith and Fisher, 1975)

Definitions
erect = to build
demolish = to knock down
legal procedures = court cases
municipality = Ankara's city government

R   What is a gecekondu?

Here are the results of a survey given to residents of a gecekondu on the edge of Ankara.

1   How long have you lived in this area?
under 10 years – 71%
over 10 years   – 29%

2   Where was your home before you came here?
another city   – 2%
another town – 17%
a village   – 78%
other   – 13%

3   Why did you choose this area?
relatives lived here – 32%
to get paid work   – 18%
rent is cheap   – 14%
other   – 36%

4   What paid work do you do?
non–manual work   – 34%
skilled manual work – 28%
unskilled work   – 22%
service work   – 16%

5   If you could change your home, what would you do?
stay if the house and area were improved   – 43%
buy our own home – 54%

6   How often do you visit the village you came from?
never   – 16%
rarely   – 42%
a few times a year   – 35%
once a month or more – 4%
I send cash regularly – 24%

7   What are your main likes and dislikes about the area?

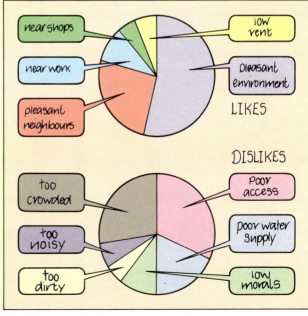

S   What are Ankara's squatters like?

A The divide and rule technique (*Fighting Apartheid*, IDAF, 1988)

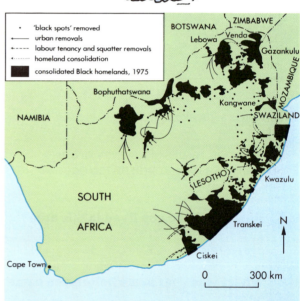

B What are the Bantustans? (*IDAF, 1988 (cartoon): Griffiths, 1984 (map)*)

C Social well-being of different groups in S. Africa (*New Internationalist, May 1986*)

	Popn (Millions) (1983)	Average monthly income (Rand) (1984)	Educucation (Expenditure p. person) (1983–4)	Housing shortage (1985)
Black	24.1	273	234	420 000
White	4.8	1834	1654	2 000
Coloured	2.8	624	569	43 000
Asian	0.9	1072	1088	18 000

1 rand = 20p (1989 exchange rate)

Type of removal	No. people (millions)
A Evicted from 'white' farmland	1.1
B Clearance from 'black spot'	0.7
C Moving people from white urban areas to Homeland township	0.7
D Removing people from illegal settlement	0.1
E Re–zoning urban land from black to white	0.8
Total	3.4

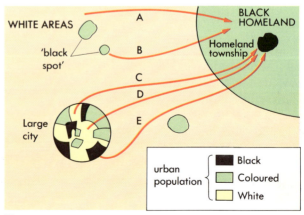

D Relocation of black population (*CIIR, 1984*)

A typical black spot clearance would involve a group of 2000 people or more whose forebears might have bought the freehold to the land as long ago as 1904. They are told that their land is a black spot and they are subject to removal. The police arrive one day to serve an eviction order which is to take effect in three months. A day before the deadline the people are told that government vehicles will be ready to move them the next day and anyone resisting removal will be arrested. Vehicles arrive under heavy police guard. Police start to remove people's belongings into the trucks and as each house is cleared it is then bulldozed. Women and children are packed into overcrowded buses, the men are left to travel on the trucks on top of the furniture for perhaps 50 km or even 300 km to a place that none of them has ever seen.

On arrival families are usually given a tent, or occasionally a corrugated iron hut, and a tin lavatory of either bucket or pit type. Water is scarce and is provided on a communal basis, sometimes with only one tap between 30 or more families. There is no domestic gas or electricity. Tents are removed after three months for the use of new arrivals. By that time it is expected that a family will have built some sort of shelter with whatever materials they find.

The land provided is too little, too dry, too rocky and too thorny to allow a family to scratch a living. No work is available on the spot and very little within commuting distances, which are often considerable. Bus fares are high and consume a large proportion of wages.

In the resettlement camps basic amenities such as shops and clinics are almost totally non-existent. Details vary from one place to another but all have poverty, hunger and ill-health in common. One survey showed that 50 per cent of all 2–3 year olds suffered from malnutrition, another 27 per cent of all 6–23 months old had kwashiorkor. In some camps the infant mortality rate is over 50 per cent – the equivalent rate among white South Africans is 1.9 per cent. Typhoid epidemics are frequent and there have been outbreaks of cholera. Tuberculosis is rife, respiratory diseases cause many deaths and when water supplies become contaminated, because of inadequate sanitary arrangements, enteritis epidemics occur.

E What does relocation mean? (*Griffiths, 1984*)

F White government workers demolishing a squatter camp

G An African woman carrying away all her wordly goods

H Typical resettlement camp

People on the move: S. Africa (2)

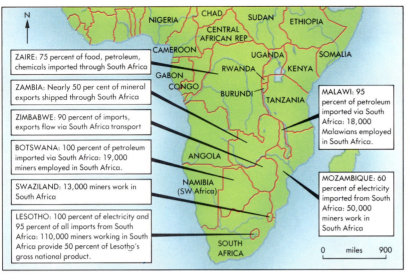

ZAIRE: 75 percent of food, petroleum, chemicals imported through South Africa

ZAMBIA: Nearly 50 per cent of mineral exports shipped through South Africa

ZIMBABWE: 90 percent of imports, exports flow via South Africa transport

BOTSWANA: 100 percent of petroleum imported via South Africa: 19,000 miners employed in South Africa.

SWAZILAND: 13,000 miners work in South Africa

LESOTHO: 100 percent of electricity and 95 percent of all imports from South Africa: 110,000 miners working in South Africa provide 50 percent of Lesotho's gross national product.

MALAWI: 95 percent of petroleum imported via South Africa: 18,000 Malawians employed in South Africa.

MOZAMBIQUE: 60 percent of electricity imported from South Africa: 50,000 miners work in South Africa

I Southern Africa's dependence on South Africa (*Guardian, 8 August 1986*)

South Africa provides many thousands of jobs for people living in neighbouring countries.

Lesotho is a small country surrounded by South Africa. The mountainous terrain and poor soils make farming difficult. Many men migrate to South Africa to find work in the mines and factories. Although wages are very low, the men earn more money than in Lesotho.

J Urban Lesotho: Maseru, the capital city

K Rural Lesotho: the Maloti Mountains

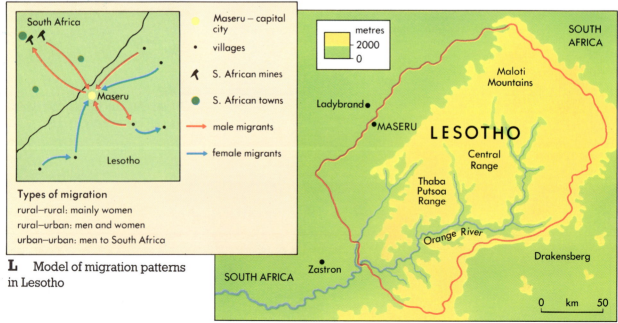

Types of migration

rural–rural: mainly women
rural–urban: men and women
urban–urban: men to South Africa

L Model of migration patterns in Lesotho

M Map of Lesotho

56

The effects of migration have an impact not only on the people who move but also on those who remain behind in the villages – the women, elderly people, and children.

N The effects of migration on women, children and the elderly (*Another Blanket, Agency for the Industrial Mission, Lesotho, 1976*)

"I go home to Lesotho about once every two years, although I regularly send money to my family."

"I think my wife has taken the children back to her mother's village in the mountains. I haven't seen them for many years."

"I wish I did not have to work in the mines, but there are no paid jobs in Lesotho. The land is too dry and barren for farming and we haven't enough land for growing crops. This is the only way I can earn enough to keep my family."

"I am Jack, the eldest son of Mr & Mrs B. I am 15 yrs old. I have a younger sister and brother. My father works in Johannesburg and my mother in Maseru. My dad comes home every 6 months and my mum every weekend. During the week I am responsible for the family. When one of the children is sick I take them to the clinic or look after the cattle. Therefore I cannot attend school regularly. This is hard for a boy of my age."

"I have been working at the gold mines in Johannesburg for almost 20 years, but there is no job security. Each 'contract' lasts about 11 months and then we have to reapply for a job 4 weeks later."

"My husband's absence has created problems. I plough my fields with great difficulties, because the person with whom we work together ploughs his fields first after the rains of spring. My field is ploughed too late."

"My son who is not well educated has been working at the mines since he started work. With the money he earns he educates his sister who is in high school. Because he does not want to spend much on transport, he only comes home at the end of his contract. When he is at work he writes and sends money home."

"I have problems with work in the fields because there is no-one to help me ploughing. Another problem is theft of cattle, crops or farming implements. Our head man has tried to stop it but it still goes on."

"My husband left me 4 years ago and has never contacted me since. I am very much worried by the difficulties I meet with my life as a human being with natural desires. I have a child with a man friend of mine, who was caring for me."

People on the move: Kampuchea

◀ **A** 'The Killing Fields'

C Map of Kampuchea

B Location of Kampuchea

58

"We set out along the crowded street, carried along by the throng. People were shouting out trying not to lose their families. Those of us who were walking held onto the car. The Khmer Rouge stood at the roadside, telling us to move along. If anyone tried to turn back, they would jump into the crowd and order him to keep moving in the general direction of the outskirts.

We moved very slowly in the heat of the day. Some people were carrying their possessions on their backs or on bicycles. Others had handcarts which they pushed and pulled. There were overloaded cyclos with families balancing on them and parents pushing. Those of us with cars were the lucky ones. Children cried out that they were being squashed in the crowd. Everywhere people were losing their relatives.

After two hours we reached a market place, where there were two piles of bodies in civilian clothes, as if two whole families had been killed, babies and all. Two pieces of hardboard stuck out of the pile, and someone had scrawled in charcoal: FOR REFUSING TO LEAVE AS THEY WERE TOLD. From here on both sides of the road were covered with dead bodies, some soldiers, some not. People were being pushed along in hospital beds. There were burnt cars and houses all around.

At the first check point the Khmer Rouge were looking for any military objects – uniforms, weapons and packs. They took some of our medicine, my watch and the Philips tape recorder and threw them on a pile of confiscated objects – televisions, cameras and cassettes. They found nothing incriminating in our car, but other groups were less fortunate. They were told to wait beside their possessions. Then, when a group of around ten had been assembled, they were led across the dry paddyfields to beyond the treeline. Then there were gunshots."

D Eyewitness account of the exodus from Phnom Penh (*Cambodian Witness*, ed. by J Fenton, 1986)

Population movement

E Population transfers, April 1975 (*Kampuchea in the 70s, Finnish Inquiry, 1982*)

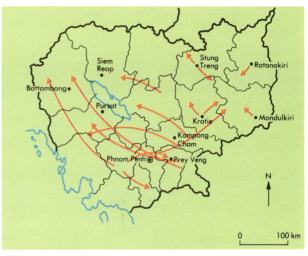

F Population transfers, May 1975–78 (*Kampuchea in the 70s, Finnish Inquiry, 1982*)

Country	Fled from Kampuchea 1	Returned to Kampuchea 2
Vietnam	150 000	130 000
Thailand	680 000	234 000
Laos	20 000	20 000
TOTAL	850 000	384 000

Resettlement provinces 3		Final destination 4	
Battambang	120 000	USA	74 200
Siem Riep	79 000	France	21 400
Prey Veng	68 000	Canada	5 800
Takeo	45 000	Australia	4 100
Svay Rieng	72 000	Switzerland	1 300
TOTAL	384 000	New Zealand	1 200
		West Germany	800
		Belgium	600
		Others	6 600
		TOTAL	116 000

G Khmer Rouge refugees, 1975–81 (*Kampuchea in the 70s, Finnish Inquiry, 1982*)

H Market busy with the daily shoppers

Life returns to Kampuchea

- December 1978; Vietnam invaded and overthrew Pol Pot and Khmer Rouge.
- Kampuchea began to recover from the nightmare 3-year rule of Khmer Rouge.
- Phnom Penh is alive with people on foot, bicycles and even a few mopeds.
- Buddhist monks are once again active in central Phnom Penh.
- Western visitors granted tourist visas in 1986. Package tours now available.
- Temples and monuments are being restored.
- Farming methods improve as life returns to normal in villages.

A Map of Belize (*Robinson and Furley, 1983*)

B Abecor Country Report, November 1986

Area:	22 965 square kilometres					
Population:	156 000 (1984)					
Gross Domestic Product per capita:	$US1178					
Independence from Britain in 1981						
Social/economic indicators:	1980	1981	1982	1983	1984	1985
GDP (% change)	+5.1	+1.6	−5.7	+2.0	+0.7	+0.5
exports ($US millions)	111	119	91	78	93	90
imports ($US millions)	136	147	116	102	118	116
national debt ($US millions)	47	56	62	75	76	94

5 year development plan
- General objective: more diversification in the economy.
- The development of export-orientated activities, notably in agriculture, forestry, fishing and tourism.
- Public investment to be directed to infrastructure projects, particularly transport, telecommunications and other services to benefit business.
- Foreign investment actively sought. Incentives to include sale of large areas of land.

C   Belize Government 5–year development plan, 1985–89

D   "Belize offers wildlife a rich variety of habitats from pine savanna on the high ridges of the Maya Mountains, tropical rainforest to mangrove swamps and limestone caves," (*Simons, 1988. This first appeared in New Scientist, London, the weekly review of science and technology.*)

Coca-Cola's development plan
Location: Hillbank area, Orange Walk District, north-west Belize.
Natural environment: Combination of dry, open pine forest and sub-tropical moist forest.
Developers: US-based transnational Coca-Cola.
Size of development: 196 000 acres
Cost of site: $US 6 million
Cost of intended investment: $US 120 million
Plan: To clear 5000 acres a year for replanting with citrus groves, eventually establishing 25 000 acres of plantation. Further land will be cleared for a processing plant and road or rail links.
Jobs: 100 jobs will be created for every 5000 acres planted. There will be additional seasonal work.

E   Coca-Cola's development

Squeezing the countryside (2)

Belize at the crossroads

Belize has some of the finest unspoilt tropical forests in the world.
Will its forests survive?

F A Jaguar in a wildlife sanctuary in Belize

Belize Government

- Our Gross Domestic Product will increase by 8% per capita per annum.
- They will build new roads and maybe even a railway.
- They will give us some land for growing crops.
- They will set up a nature reserve.
- Belize Zoo will get land for breeding tapirs.
- It will mean jobs for a lot of our unemployed.

Coca-Cola's point of view

- We can't rely on Florida citrus groves any more – frost and disease have killed them off.
- It is frost-free in Belize.
- Belize is a peaceful country.
- When we bought the land in Belize it was very cheap.
- There is a large supply of labour and it is very cheap.
- We can get cut-price tariffs for exporting fruit and concentrate.
- Lots of our land in Belize is suitable for orange growing.
- Our efforts will help Belize on the road to economic success.
- We CARE about the environment and will give generously to help its conservation.

G Coca-Cola's development: the arguments for

Coca-Cola hasn't listened to us.

What happens if Coca Cola doesn't stay in Belize?

Thousands of wildlife species will be lost.

Already miles of forest have disappeared.

What about the howler monkey and the tapir?

Mangroves have been removed to make sandy beaches for tourists.

The Audubon Society
WHO ARE THEY?
- It is the local environment group in Belize.
- Founded in 1969 to raise funds to protect bird life. Today it runs five government wildlife preserves.
- It has the ear of the Government.

I The Audubon Society

We are not against controlled economic development.

Where are the safeguards? the whole business stinks.

The ecosystem is very fragile.

Just another case of exploitation

Risk of soil erosion and water pollution from pesticides and fertilizers.

Small citrus growers already established will be out of business.

Quality of life for people living in area area around Belize City will be affected.

Coca-Cola is in a strong position to help Belize.

They could help preserve the natural habitat.

FOR / AGAINST

THE COKE CONNECTION

H Coca-Cola's development: the arguments against

Friends of the Earth

TURN YOURSELF INTO A FIRE EXTINGUISHER.

The burning of the rainforest is currently the greatest environmental folly on Earth.

Last year an area the size of Belgium was torched. This year promises to be as bad, with ever more CO_2 being pumped into the atmosphere, aggravating the Greenhouse Effect.

But Friends of the Earth can actively help prevent such devastation by putting pressure on governments and companies.

There's no earthly reason not to join us, and help extinguish the fires.

THE EARTH NEEDS ALL THE FRIENDS IT CAN GET

J Friends of the Earth poster, 1989

Reactions to Coca-Cola's pullout were bitter. The government was outraged that foreign pressure had forced the company out, particularly as the foreign press that ran the story from the Friends of the Earth did not consult the ministers concerned.

Minister Lindo says: "I feel that people like Friends of the Earth should mind their own business." The government now fears that the collapse of the Coca-Cola project has jeopardised the attractions of the country to other foreign investors.

K Stop press: Coca-Cola pulls out of Belize (*Simons, 1988*)

Acknowledgements

Thanks are due to the following for permission to reproduce photographs:

Ace Photo Agency pp. 12(C), 13(F); Aviemore Photographic pp. 41(D, G and H), 44(M), 46(S); Barnaby's Picture Library p. 12(D); The Bridgeman Art Library p. 4(A); Cambridge University Collection of Air Photographs p. 43(J); Camera Press/Stefan Sonderling p. 55(H); J. Allan Cash Ltd. pp. 4(B and F), 5(I), 9(U), 40(reindeer), 49(F and G), 57(young boy and women in fields); Council for the Protection of Rural England p. 25(G); Greg Evans Picture Library pp. 5(K and L), 7(O); Mary Evans Picture Library p. 4(E); Rex Features p. 55(F and G); Goldwater/Network p. 13(E); David Gowans p. 45 (O and P); Robert Harding Picture Library pp. 4 (C and D), 11(F), 20(F), 36(F), 40(conifers), 45(Q), 49(H), 61(Belize forest and rainforest); Hutchison Picture Library p. 5(H and J), 51(M), 56(J and K), 57 (mineworker); The Kobal Collection p. 58(A); Network Photographs p. 36(C); NHPA p. 20 (badger run); Oxford Scientific Films pp. 20 (badger), 40(B), 40(moss campion, bilberry heather and grouse), 61(mangrove swamp), 62(F); Chris Ridgers p. 7(N); John Sturrock/Network p. 5(G); *Watford Observer* p. 23(M); Kerry Wills pp. 30(B), 38(C and D); C. R. Wood p. 30(A); University of Cambridge p. 47(W).

Cover photograph by Tony Stone, Worldwide.

We are grateful to the following for permission to reproduce copyright material:

Agency for the Industrial Mission, Lesotho for extract N on p. 57 from *Another Blanket*, 1976; Berkshire County Council for map D on p. 31; Bracknell District Council for extracts A–E on pp. 34–35; Cambridge University Press for map L on p. 17 from R. Hall and P. Ogden *Europe's Population in the 1970s and 1980s*, 1985; Carngorm Chairlift Company Ltd for plan R on p. 46; CIIR for table D on p. 54; Consortium Development Ltd for figures E and F on P. 24, extract H on p. 25, diagrams J–P on pp. 26–27 and the press release extracts on pp. 24 and 26; The Controller of Her Majesty's Stationery Office for extract S (*Regional Trends*, No. 23, 1988) on p. 9, graph T (*Social Trends*, No. 18, 1988) on p. 9, table I (*Population Trends*, No. 27, Spring 1982) on p. 16, extract B on p. 18 from Government circular PPG2, *Planning Policy Guidance: Green Belts*, Department of the Environment; Council for the Protection of Rural England for extract G on p. 11; Countryside Commission for table B and extract D on p. 10 from *Household Surveys for the Countryside Commission*, Summer 1986 and for illustration C from *Countryside Commission News*, 18, December 1985; *Daily Telegraph* for map K on p. 22 and extract L on p. 23; Department of Transport for map I on p. 21 from *The M25 Orbital Motorway*, 1986; Elsevier Science Publishers Ltd for the table on p. 9 from *Rural Deprivation and Planning* by C. Thomas and J. Winyard; *Esso Magazine* for 'M25 – the Mighty Motorway' on p. 20; Faber and Faber Ltd for extract D on p. 58 from *Cambodian Witness: The Autobiography of Someth May* ed. by J. Fenton; Finnish Inquiry Commission for maps E and F and table G on p. 59

from *Kampuchea in the 70s*, 1982; Fishers of Keswick for postcard T on p. 46; Ford Foundation for table C on p. 49; Friends of the Earth for figure J on p. 63; The Geographical Association for table K on p. 36 from Y. K. Court 'Recent patterns of population change in Denmark', *Geography* 70(4) (1985) p. 354, maps A and C on p. 18 from R. Munton 'Green Belts: the end of an era?' *Geography* 71(3) (1986) p. 208 and map A on p. 60 from Robinson & Furley 'An independent Belize', *Geography* 68(1) (1983) p. 44; *Geographical Magazine* for figure C on pp. 14–15 based on 'Millionaire cities of the seventies' (using revised data) of May 1978, map A on p. 40 of March 1980, extract L on p. 44 of March 1980, map B on p. 54 and extract E on p. 5, both from I. Griffiths 'When a Homeland is not a Homeland', December 1984; *Guardian* for map I on p. 56; International Defence and Aid Fund for cartoons A and B on p. 54 from *Fighting Apartheid*, 1988; Mary Glasgow Publications Ltd for table H on p. 16 from *Geogile* (29), January 1984; Frank Herholdt (photographer) for the MKDC for figure J on p. 16; Michelin for extract A on pp. 12–13 from map no. 55; New Internationalist for the data in table C on p. 54 which appeared in the May 1986 edition (data supplied to NI by South Africa Institute of Race Relations, Market Research Africa and *Weekly Mail*); New Scientist for caption D on p. 61, heading and introductory text on p. 62 and extract K on p. 63 all from P. Simons 'Belize at the Crossroads', *New Scientist* 29, October 1988; Open University Educational Enterprises Ltd for extract D on p. 49, maps J and K on p. 50 and the quotation in extract R on p. 53, all from the 'Third World Studies Series', *Migration: The Turkish Case*, 1983 and for extract N on p. 51 from *Sakaltutan – a time of change*; Ordnance Survey (crown copyright reserved) for extract D on p. 19 from 1:25000 Pathfinder 206, extract I on p. 25 based on 1: 25000 Routemaster 9, extract K on p. 27 from 1:10000 map, extract B on p. 28 from 1:250000 Routemaster 9, extract A on p. 32 from 1:25000 Pathfinder SU87/97 and 86/96, extract B from 1:50000 Landranger 174; Oxford University Press for diagram E on p. 49; Regional Studies Association for Map Q on p. 8; Routledge for map A, graph D and chart E on p. 36, maps G and H on p. 37 and maps A and B on p. 38, all from J. K. Bowers and P. Cheshire *Agriculture, the countryside and land use* (Methuen & Co, 1983); Spey Valley Tourist Organization for extract F on p. 41; Thames Television for extract E on p. 39; Town and City Properties (Development) Ltd for map J on p. 22; University of Durham (Department of Geography) for maps O and Q on p. 52 and the quotation in extract R on p. 53 from D. W. Drakakis-Smith and W. B. Fisher 'Housing Problems in Ankara'; *Watford Observer* for extract M on p. 23; John Wiley and Sons for the quotation in extract R on p. 63 from G. Payne *Low Income Housing in the Third World*, 1984; Kevin Woodcock for cartoon B on p. 24 published in *Private Eye*; Worldwide Fund for Nature for figure E on p. 11 (drawn by Mike Bilsland).

Every effort has been made by the publishers to obtain permission for reproducing copyright materials but if any material is incorrectly attributed, the publishers would be happy to hear from the correct copyright holder.